The Traveller's Guide to MARS

ALSO AVAILABLE IN THE TRAVELLER'S GUIDE SERIES

The Traveller's Guide to The Centre of the Earth

The Traveller's Guide to The Big Bang

The Traveller's Guide to Infinity

This edition first published in 2018 by
Palazzo Editions Ltd
15 Church Road
London SW13 9HE

www.palazzoeditions.com

A CIP catalogue record for this book is available
from the British Library.
ISBN: 978-1-78675-062-4
Printed and bound in China

Created by Hugh Barker for Palazzo Editions Ltd
Cover art and illustrations by Diane Law

The Traveller's Guide to MARS

Colin Stuart

PALAZZO

ABOUT THE AUTHOR

Colin Stuart is an astronomy author and speaker who has talked to over a third of a million people about the universe. His books have sold more than 150,000 copies worldwide and been translated into eight languages. He's written over 150 popular science articles for publications including The *Guardian*, *New Scientist*, *BBC Focus* and the European Space Agency.

In recognition of his efforts to popularize astronomy, the asteroid (15347) Colinstuart is named after him and in 2014 he was awarded runner-up in the European Astronomy Journalism Prize. A Fellow of the Royal Astronomical Society, he's talked about the wonders of the universe on Sky News, BBC News and BBC Radio 5Live

Contents

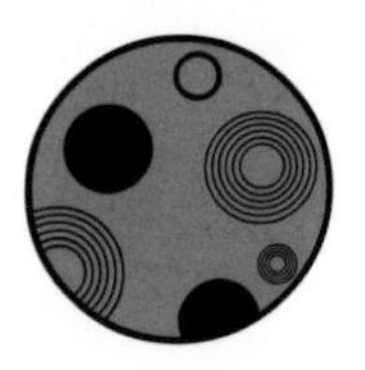

Introduction

Mars has fired our imaginations more than any other planet. It has captivated, enthralled, and intrigued us for centuries. Our curiosity has sparked an armada of robotic missions to the Red Planet, which have sent back stunning images of sweeping plains, ancient volcanoes, and sunsets on another world.

But it's not enough. We humans have a natural sense of adventure and a need to explore. Having conquered the moon in the 20th century, all eyes are on Mars in the 21st. It seems increasingly likely that before 2100 – perhaps much sooner – the first crew will be stepping out onto the red dust.

Trade and tourist routes will follow, just as Europeans flocked to the New World after the first explorers reached it. People will routinely sell all their earthly possessions in exchange for a one-way ticket costing the same as a middle-class house in the West.

What will this colony be like? Consider this book your travel guide, a *Lonely Planet* for the Red Planet. Much of what follows is established scientific fact, based on the latest findings from the Mars rovers and the best theories of planetary scientists. But I've also used what we know about Mars to imagine a future world which has a permanent human presence.

Full speed ahead!

This image of Mars shows the Valles Marineris hemisphere of the planet. (The Valles Marineris is the slanted streak, a significant canyon system). This is the view you would expect to see from a spacecraft approaching the Red Planet.

Map and Fact File

Diameter: 6,779 kilometers (4,212 miles): 53% of Earth's
Mass: 0.64 trillion trillion kilograms (1.28 trillion trillion pounds): 10.7% of Earth's
Moons: 2 (Phobos and Deimos)

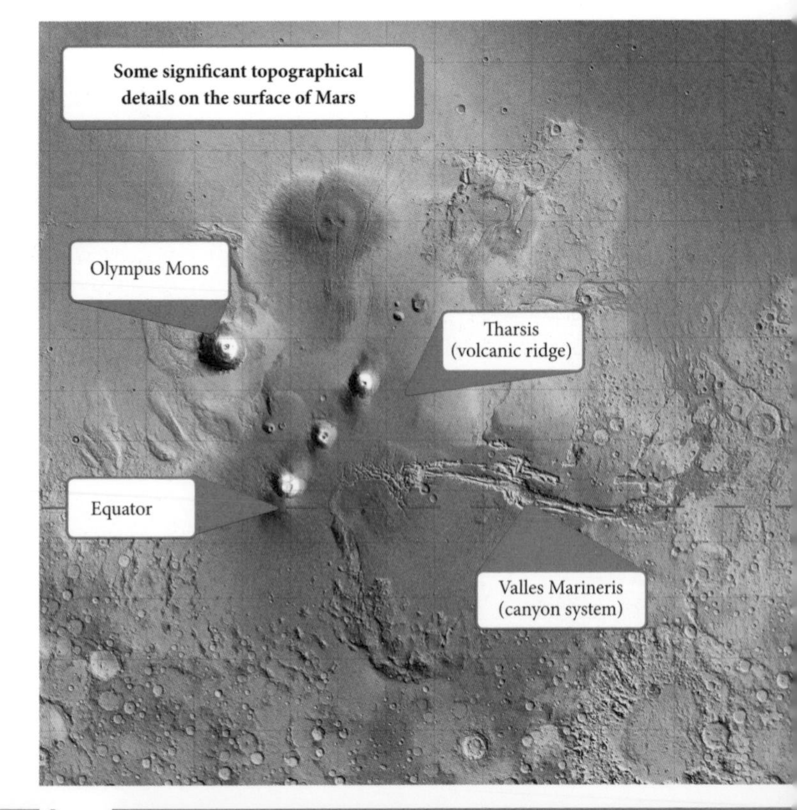

Average distance from the sun: 228 million kilometers (142 million miles)
Orbital Period: 1.88 Earth years
Day: 24hr 37 minutes 22 seconds
Surface gravity: 3.7 m/s^2 (4.05 yards/s^2) = 37.6% of Earth's
Escape velocity: 5.03 km/s (3.13 miles/s) = 44.9% of Earth's
Atmospheric Composition: 95.97% carbon dioxide, 1.93% argon, 1.89% nitrogen, 0.15% oxygen 0.06% carbon monoxide
Atmospheric Pressure: 0.00628 atm (Earth = 1 atm)

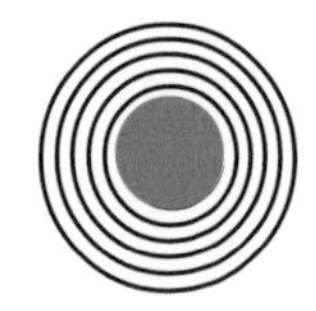

The Solar System

The region occupied by the known planets is about 9 billion kilometers (5.6 billion miles) in diameter. The closest planet to the Sun is Mercury, followed by Venus, Earth, Mars, Jupiter, Saturn, Uranus, and Neptune. Beyond Neptune is Pluto, the first object to be found in the Kuiper Belt, which was originally considered to be the ninth planet but has now been reclassified as a dwarf planet.

Mars is a minimum of 54 million kilometers (34 million miles) from the Earth, which is just over 150 million kilometers (93 million miles) from the sun, around which all the planets are in rotation.

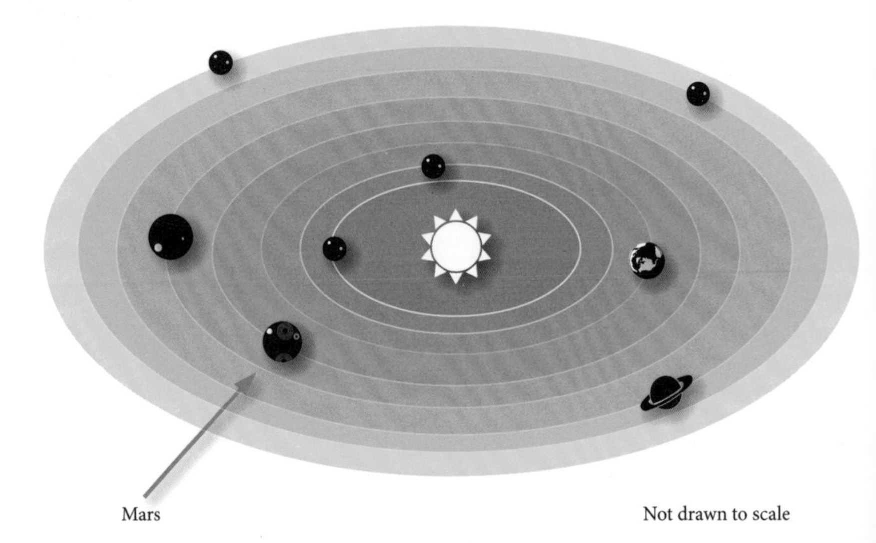

Weather & Climate

Climate is what we expect.
Weather is what we get.

Mark Twain

Seasons

On Earth we experience a repeating pattern of four seasons that have the same intensity across both hemispheres. Winter in Scotland isn't that different from winter in Patagonia. That's because our distance from the sun barely varies across one orbit.

Mars, on the other hand, has the second most eccentric orbit of any planet in the solar system. That means the shape of its path round the sun is far from circular (see diagram). It can get as

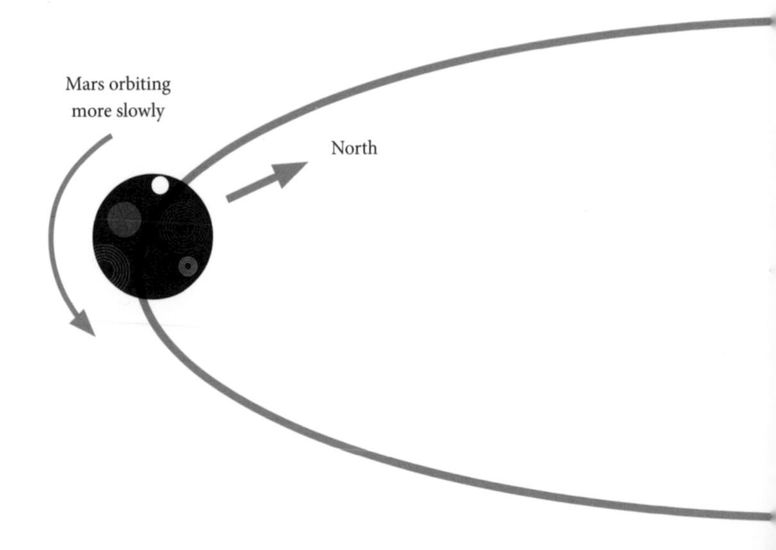

close as 206 million kilometers (128 million miles) and as far as 249 million kilometers (155 million miles). The nearest point is known as *perihelion* while the furthest point is called *aphelion*.

Summer in the northern hemisphere – when the north pole of Mars is tipped towards the sun – coincides with aphelion. By contrast, summer in the southern hemisphere occurs at perihelion. So Southern summers are a lot shorter and hotter than Northern summers, and Southern winters are longer and colder than those in the north. Mars's axis is currently tipped over by 25 degrees, which is very close to Earth's axial tilt of 23.4 degrees. However, without a large moon for ballast, Mars is significantly affected by the gravitational pulls of the other planets. This can cause it to be tipped over by as much as 45 degrees.

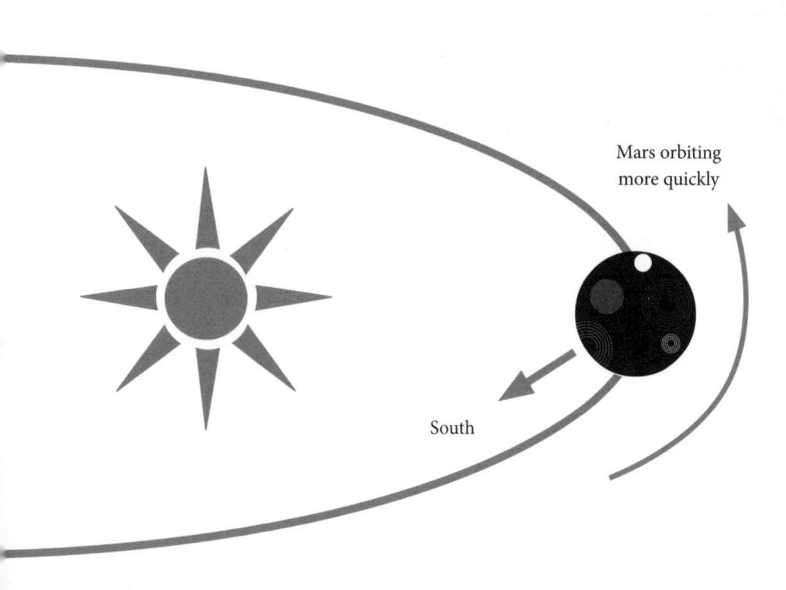

The Weather Forecast

First the good news: with a day that's 24 hours, 37 minutes and 35 seconds long, the planetary rotation of Mars is the closest to Earth of any planet in the solar system. Once on the surface, your body clock won't be thrown too much out of whack. Do bear this time difference in mind though, if you are trying to keep in touch with loved ones back home.

After a ninety day stay you'll be more than two days out of sync. You'll have to get used to considerably lower light levels, too. Out on Mars, the sunlight is on average only around 40% as intense as the sunlight we receive on Earth.

This often results in particularly frigid temperatures because the thin atmosphere of Mars isn't good at transporting heat around the planet. Nor is its famous ruddy soil efficient at storing solar energy.Temperatures can reach a balmy 35 degrees Celsius (95 degrees Fahrenheit) at the equator during Martian summer, but they can also plummet to an inhospitable -143 degrees Celsius (-225 degrees Fahrenheit) at the poles in winter.

With average temperatures at a frosty -63 degrees Celsius (-81 degrees Fahrenheit) you'll be well advised to stick to your accommodation and only venture outside with the correct protective equipment.

Getting There

It's a fixer-upper of a planet but we could make it work

Elon Musk

Medical

A trip to Mars is not your average holiday. Such a daring vacation comes with a significant amount of medical baggage and you'll have to be in decent shape to get the green light to travel.

You'll need to be free of diseases, even minor ones, and substance addictions are an obvious no-no. A history of psychiatric disorders will likely see your travel clearance revoked. The standard test is NASA's long-duration astronaut physical examination. Your blood pressure cannot exceed 140/90 in a sitting position, and your eyesight needs to be correctable to 20/20. This particular restriction may be relaxed once a thriving colony is established and visitors are genuinely being treated as tourists.

Height and weight are crucial, too – spacecraft are small and, when it comes to space travel, extra weight brings a hefty cost premium. Applications from those shorter than 1.57m (five feet two inches) or taller than 1.91m (six feet two inches) are unlikely to be accepted, at least in the early days.

You must be able to swim 75 metres (246 feet) without stopping, and then swim the same distance in a flight suit and tennis shoes with no time limit. Still wearing your flight suit, you'll need to be able to tread water continuously for ten minutes.

Astronaut Akihiko Hoshide of the Japan Aerospace Exploration Agency (JAXA) wears a training spacesuit in the water at the Neutral Buoyancy Laboratory near NASA's Johnson Space Center

Astronaut Training

Before you depart for the Red Planet, you will be obliged to undergo a series of training exercises to make sure you are fully prepared for the rigors ahead. A wide variety of technical skills are necessary, from engineering to the basics of medicine and first aid. Training in geology is also provided to allow you to understand the terrain on Mars, as you will be expected to keep an eye out for unusual items of potential scientific value.

During the mission you'll be isolated in a way you've never experienced before and therefore resilience training is crucial. You'll be schooled in ways to combat home-sickness and melancholy. Effective teamwork is also a must and can be particularly difficult at such close quarters. A lot of your training will be spent on perfecting collaborative projects and conflict resolution. Fully immersive mock-ups of the Martian surface will make these as realistic as possible.

You'll train in a centrifuge, to teach you how to cope with the high acceleration of launch, and might even take a trip on the fabled "vomit comet" – an aeroplane flying on parabolic loops to simulate weightlessness in space. Once you've passed your astronaut training you will be eligible for a place on a launch vehicle and ready to depart on the next available flight.

Astronauts training for the experience of weightlessness on the "vomit comet."

Launch

You can't just go to Mars whenever you want. Launch opportunities are restricted to short windows every 26 months, at a point when Earth and Mars are favorably positioned on the same side of the sun.

Crews are lofted into a parking orbit and await everyone else's arrival before hundreds of ships depart as one armada to the Red Planet. Each spaceship has room for around 100 passengers.

Standing on Launch Pad 39A at NASA's Kennedy Space Center in Florida, your spaceship will be strapped to the most powerful rocket ever built. Standing a whopping 106 metres tall, SpaceX's *Big F****** Rocket* (*BFR* – and yes, that's really what SpaceX call it) is powered by 31 Raptor engines. With 5.4 million kilograms (12 million pounds) of thrust, it has enough power to deliver 150-250 tonnes to a low orbit around the Earth.

Once your ship has detached, the rocket booster will fall back to Earth and autonomously land on SpaceX's floating drone ship *Of Course I Still Love You* which will be positioned in the Atlantic Ocean. It will be transported back to Florida to be reused.

You'll then be free to undo your seatbelt, experience weightlessness and drink in the spectacular view of Earth from above. Once the armada is complete, your spaceship's six engines will fire and blast you on your way to Mars.

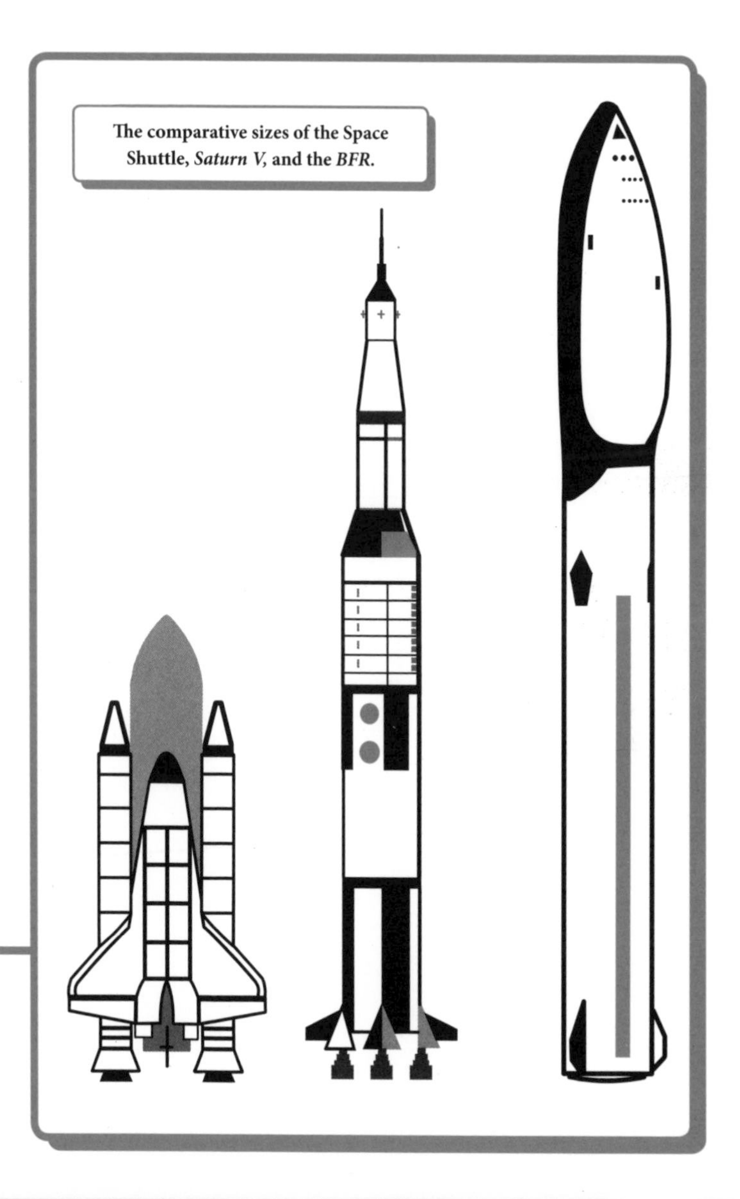

The comparative sizes of the Space Shuttle, *Saturn V*, and the *BFR*.

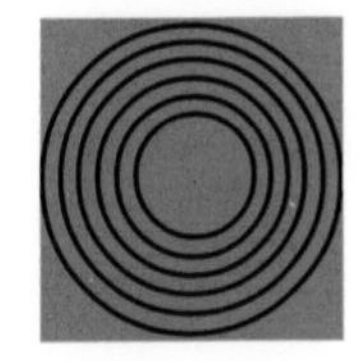

The Journey

The *BFR* spaceship, your home for several months en route to Mars, is surprisingly roomy. It is 48 metres (157 feet) in length and 9 metres (30 feet) in diameter, including the propellant tanks and rear engines. The payload area at the front totals 825 cubic metres (318 cubic feet) – larger than the inside of an Airbus A380 aircraft – and is pressurized for human habitation. It is divided into 40 cabins, with 2-3 people living in each one. There's a galley for food preparation and large communal areas for entertainment and storage, along with a solar storm shelter.

Behind you is a carbon fibre fuel tank with room for 1100 tonnes of propellant that will come from a refuelling ship in orbit to minimize the initial launch mass of your rocket. It will power four vacuum engines and two sea-level engines that together will accelerate the ship to speeds of 100,000 kilometers per hour.

A single delta wing – triangular in shape and named for its resemblance to the Greek letter delta (Δ) – will help control the pitch and roll of the ship as it approaches the Martian atmosphere.

The same ship will bring you home, having refuelled on Mars by combining water with carbon dioxide in the Martian atmosphere to create methane and oxygen rocket fuel.

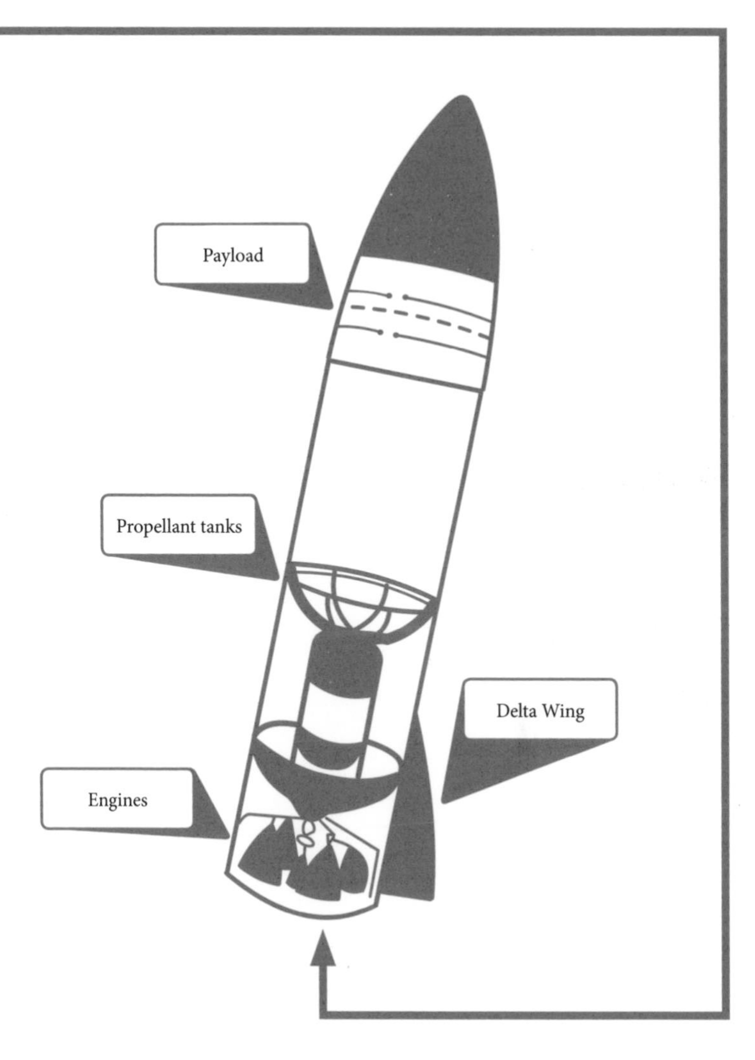

Diagram of one potential design for the *BFR*

Staying Fit

One day, a trip to Mars might take place on a rotating ship that mimics the effect of Earth's gravity. However, in the early phase of travel to Mars, you are going to be largely weightless during your mission and weightlessness is not without biological consequences. As they are no longer supporting your bulk, your muscles and bones will begin to degrade. You can lose up to 30% of your muscle performance and up to 15% of your muscle mass. Your bones would weaken as much in a three month period as they do in a decade on Earth.

To avoid these pitfalls you will need to stick to a highly regimented exercise regime that lasts for between two and four hours a day. These exercises have been tried and tested on the International Space Station. Specially adapted treadmills, exercise bikes and weightlifting equipment have been fitted to a small gym in the communal area of your spaceship. A strict roster will be in place to make sure that all crew members get their allotted daily exercise.

Wearable devices – which must be worn at all times – will track your vital signs and constantly update mission control back home about any warnings that health issues are imminent.

Astronaut Tim Peake using the muscle measurement machine MARES aboard the International Space Station.

Seat

In-Flight Meals

Space food has a poor reputation. When NASA astronaut John Glenn became the first person to eat away from the Earth, he had to make do with puréed apple sauce. I'm afraid your diet en route to Mars isn't going to be much better. Space during the journey is at a premium. While there is a galley, it doesn't have enough room for an elaborate larder stocked with reams of cooking equipment.

Weightlessness affects food as well as people – packaged food floating in the galley area of the Zvezda Service Module on board the ISS.

A hydroponics warehouse in Brooks, Alberta

All food consumed during transit is loaded before the launch, so items with shelf lives of at least seven months are necessary. Get used to eating dense food bars that are packed full of nutrients and contain up to 700 calories each.

However, psychological studies have shown the benefits of the occasional culinary treat. So fresh meals are prepared twice a week using ingredients from a small hydroponic farm on board with a limited range of fresh plants that are easy to grow. Be prepared for large quantities of lettuce, for instance.

Thankfully, the range of food available is more varied once you arrive on Mars as space is at far less of a premium on the surface of the planet. Your habitation has a dedicated team of chefs and menus are designed to make the most of the precious food grown there.

Your Spacesuit

During your journey you'll be free to roam about the spaceship without worrying about pressure or oxygen. However, as you start to approach Mars, you'll be required to don your pressurized spacesuit in preparation for landing. The suit must also be worn at all times while on the Martian surface. The low surface pressure on Mars means that the warmth of your body would be enough to boil your blood.

Gone are the days of the bulky, restrictive spacesuits worn by the *Apollo* astronauts and International Space Station spacewalkers. Instead, you'll be issued with your own skintight BioSuit designed by researchers at the Massachusetts Institute of Technology (MIT). It provides the necessary pressure by pressing directly on your skin through a series of interconnected nickel-titanium coils that contract to perfectly fit your shape once it is plugged into a power supply. Cooling the suit relaxes the coils, allowing it to be easily removed. It has the added advantage that any incidental damage can be quickly and easily repaired with special bandages.

You'll also have a helmet with an oxygen supply, communications array, and a tinted visor that will protect you from the rays of the sun. The visor is augmented with a visual display that overlays vital information about your environment and oxygen and radiation levels.

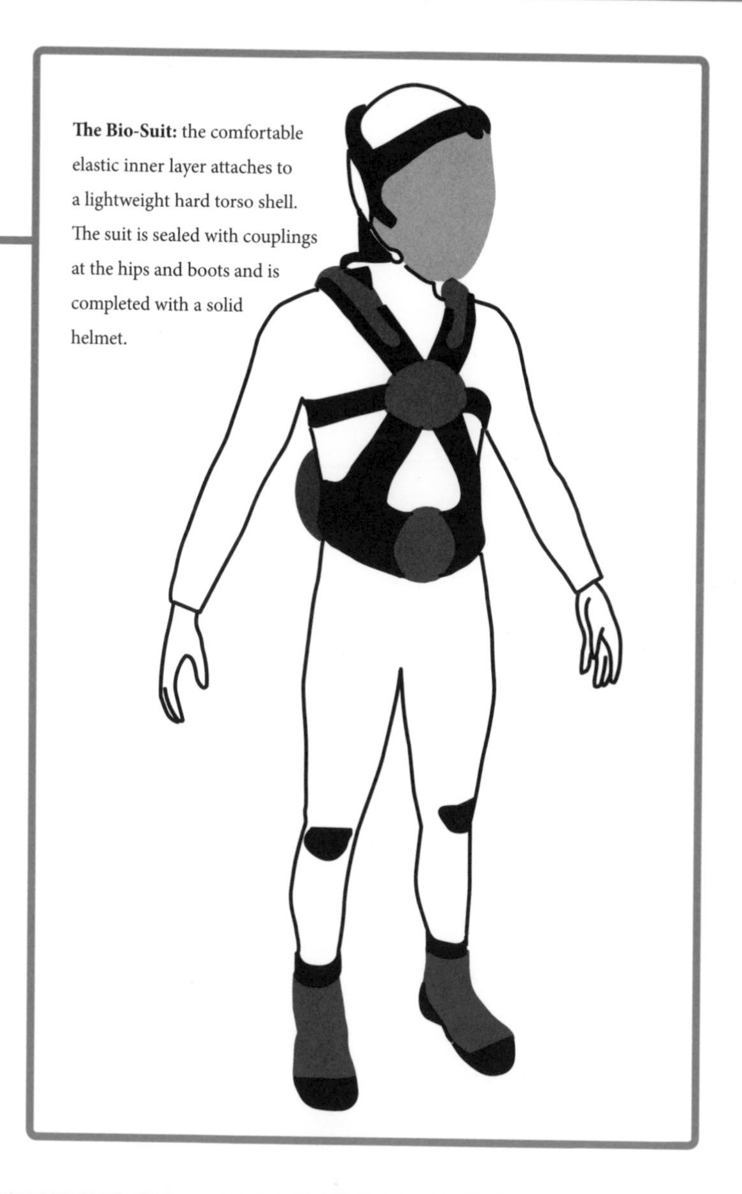

The Bio-Suit: the comfortable elastic inner layer attaches to a lightweight hard torso shell. The suit is sealed with couplings at the hips and boots and is completed with a solid helmet.

Touchdown

Getting from the orbit of Mars to the surface of the Red Planet is no mean feat. In fact, it is hair-raisingly dangerous. Many of our early attempts to land robots on Mars ended in crashes. After NASA successfully deposited the *Curiosity* rover there in 2012, its final approach to landing was remembered as "The Seven Minutes of Terror."

The biggest problem is finding a successful way to slow down. Mars is a big enough planet for its gravity to pull you in powerfully. However, without a substantial atmosphere there is no easy way to brake. Friction with what little gas there is will heat the outside of your ship to 1,700 degrees Celsius (3,100 degrees Fahreheit) as you descend.

Robotic missions have relied on parachutes, airbags, and sky cranes to slow down, but none of those will be able to halt the progress of such a massive spacecraft. Instead, supersonic retro-propulsion boosters will fire, then landing feet will be deployed immediately before touchdown.

Giving yourself the greatest distance to brake is the best way of ensuring a better chance of success, so we've built the spaceport on Mars at a low elevation. You'll land at the western end of the Valles Marineris, ready to transfer through the Noctis Labyrinthus (a maze-like system of deep valleys with steep walls) to your accommodation at Pavonis Mons.

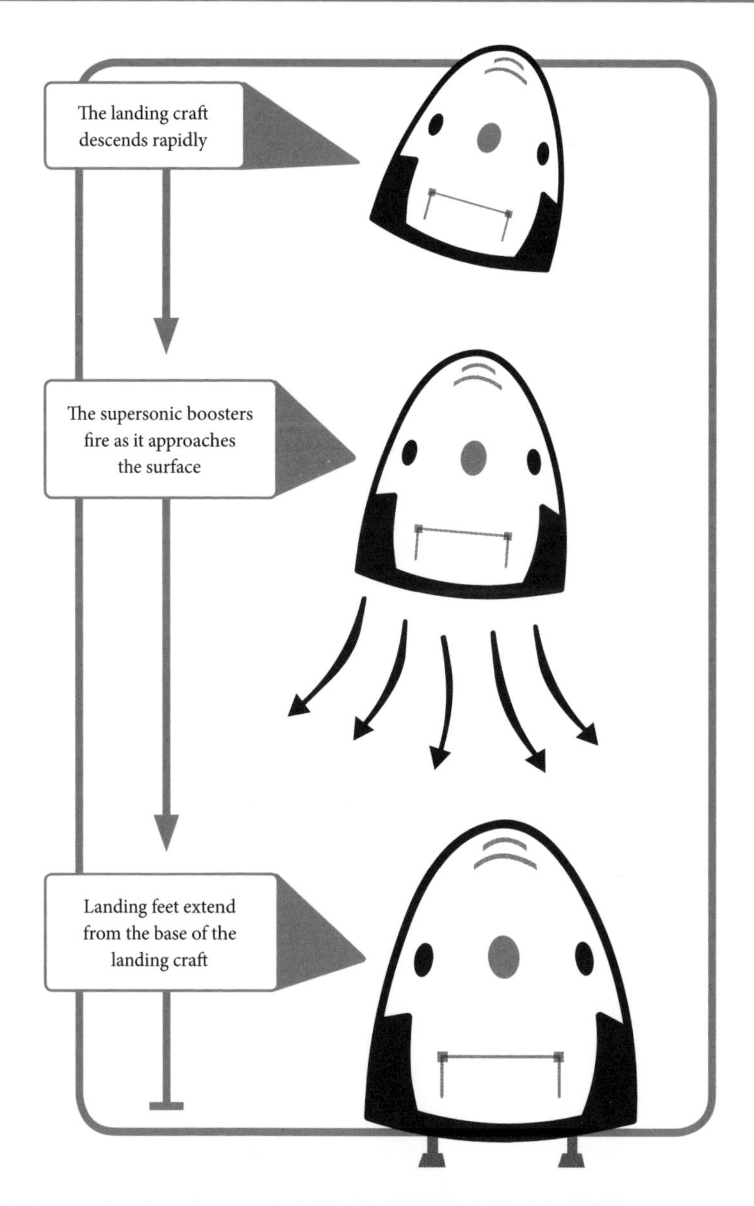
The landing craft descends rapidly
The supersonic boosters fire as it approaches the surface
Landing feet extend from the base of the landing craft

Your Accommodation

The myriad dangers of Mars, which include meteorites, intense ultraviolet radiation, dust storms, and solar flares, mean that you'll be taking refuge in the natural shelters offered by the planet's geology. While a grander surface colony is currently under construction, you can expect your home to be a makeshift hotel built into the volcanic caves nestled under the flanks of the Pavonis

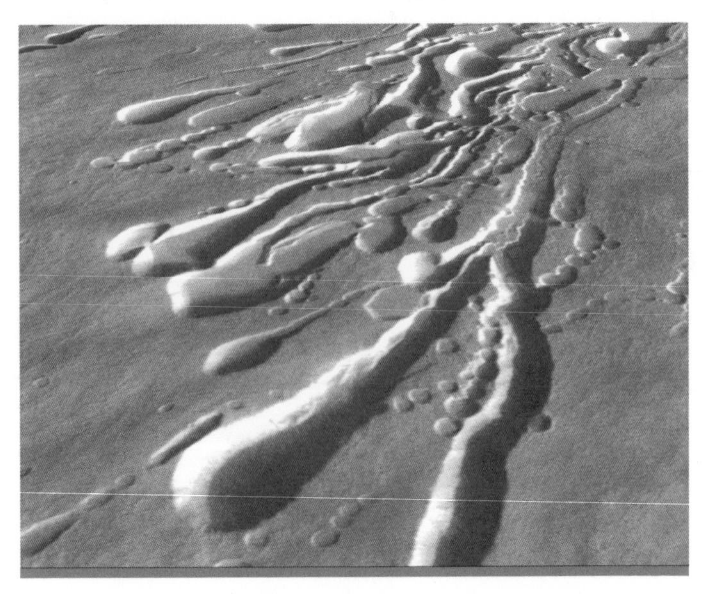

The lava tubes of Pavonis Mons photographed with the High Resolution Stereo Camera (HRSC) on board ESA's *Mars Express*

Artist's impression of a spacesuited explorer on Mars

Mons volcano. It's ideally located, close to the planet's main tourist attractions, including Olympus Mons and the Valles Marineris.

Formed by ancient lava flows beneath the Tharsis bulge, these roomy caverns provide ideal protection from the ravages of the hostile environment on Mars. The cave roofs can be as high as ten metres (33 feet). Conveniently, temperatures below ground fluctuate a lot less than on the surface.

A series of natural skylights allow daylight to flood down from the Martian surface. In 2010, students from Evergreen Middle School in California spotted a 190 metre x 160 metre (623 feet x 524 feet) skylight in NASA images of the area.

Dust Storms

Have you ever returned home from a beach holiday with sand still in your shoes? That's nothing compared to the dust on Mars. It gets *everywhere*. Dust storms as big as continents can rage for weeks, covering everything in a thin blanket of red particles. Localized storms are common, but they pale in comparison to the global dust storms that occur roughly every three Martian years (and which can lead to all activities being cancelled as visitors hunker down and wait for the worst to subside). The largest storms happen during summer in the southern hemisphere. Air

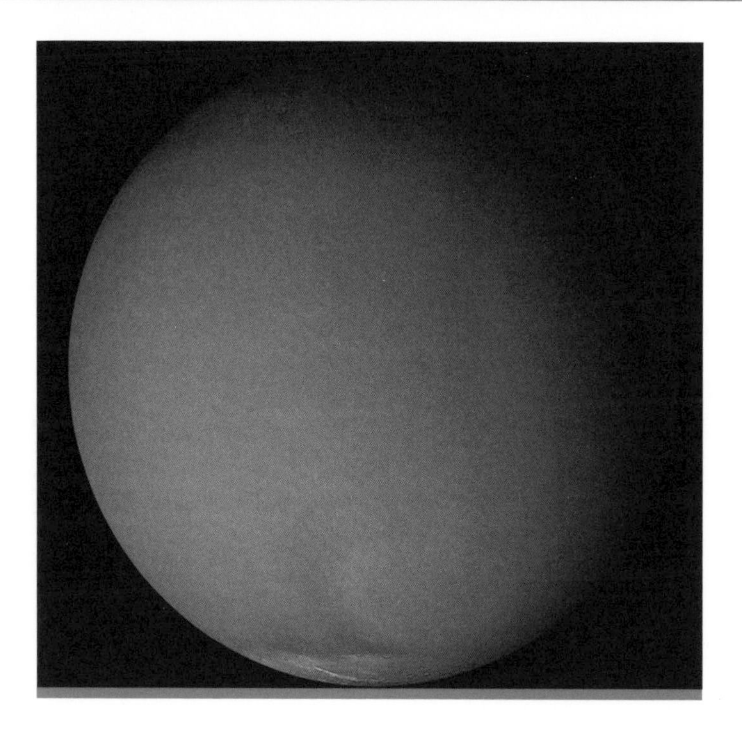

rushes from hot regions to cooler ones, creating fierce winds that whip up dust from the dry Martian surface. Temperatures and light levels drop as dust blocks out the sun. Solar panels are caked in a fine powder, further reducing the ability to generate electricity. In 2007, the *Spirit* and *Opportunity* rovers had to go into hibernation during a global dust storm to avoid being damaged. Unlike on Earth, there is no rain to wash the dust out of the air, so the atmosphere remains thick with particles for much longer. The "before" and "after" images above, taken from the *MGS Mars Orbiter* Camera show the dramatic impact of a 2001 dust storm that enveloped the planet.

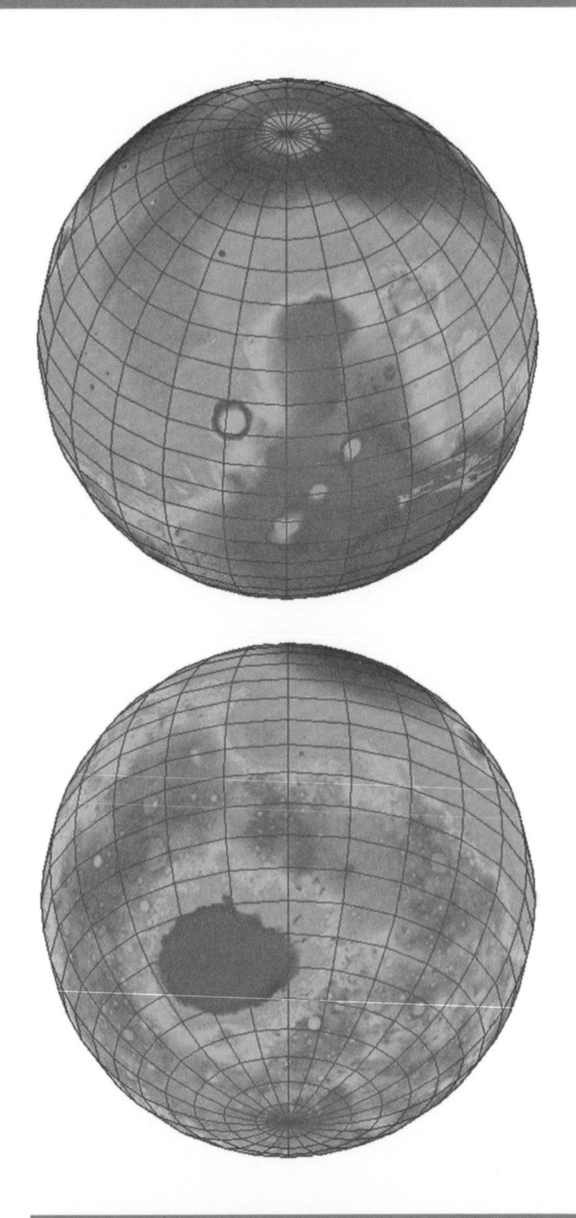

Radiation

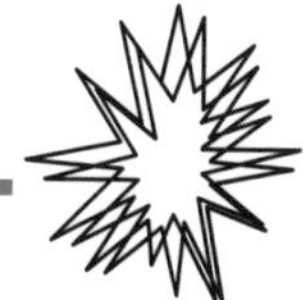

One major consideration for any out of this world trip is that your fragile body was very much engineered by evolution to be able to survive on this particular planet. Venture away from its protective environment and all of a sudden you'll face risks that you've never even had to think about before.

Chief among these considerations is the tissue-damaging effect of radiation from space. Here on the ground you are shielded from such worries both by our planet's atmosphere and its magnetic field, which together act as a giant bubble keeping out the worst that the cosmos has to throw at us. However, a trip to Mars is not so cosy.

In transit, and on the Martian surface itself, you'll be subjected to high energy particles from the sun and from exploding stars in the wider galaxy at large. Left unprotected on the Red Planet you will be zapped with more radiation than a full body CT scan every single week. So, as fun as Mars may be, and as much as your hosts will try to protect you from these dangers, prolonged stays are not advisable. In short, pack as much in as you can and limit your time away from the protective pods that serve as your accommodation.

The cosmic ray environment measured on a scale of dose equivalent values (rem/yr). The darker areas are the ones with the highest levels of radiation.

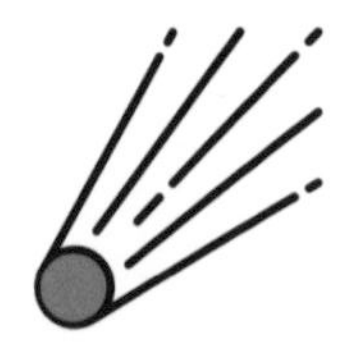

Meteorites

A thinner atmosphere also means less protection from the bevy of impactors that bombard the planets. The proximity of Mars to the asteroid belt and zodiacal cloud means that it is hit 200 times more often than the Earth and three times more often than the moon. In 1967, the *Mariner 4* probe was damaged after it was peppered with lots of tiny impacts from micrometeorites.

Two hundred objects bigger than one metre across are thought to smack into the Red Planet every year. In 2014, NASA's *Curiosity* rover stumbled upon an iron-nickel meteorite of this size. Scientists named it *Lebanon* because its outline resembles that of the Middle-Eastern country (see the photo on the right).

Another study has suggested that every three years Mars is blind-sided by an impact equivalent to one megaton of TNT. That's the same explosive power as a million tonnes of dynamite. For comparison, the biggest thermonuclear warhead ever detonated – the

The circled areas are high-resolution images taken by the Remote Micro-Imager (RMI) in *Curiosity's* Chemistry and Camera instrument. These have been combined with lower quality images taken from the Mast Camera. Iron meteorites are often found on Earth, but stony meteorites are more common. On Mars, iron meteorites are the most frequently found type, possibly because the iron is resistant to erosion processes.

USSR's Tsar Bomba – had a yield equivalent to 50 megatons of TNT.

You'd be very unlucky to get hit by one of these impacts. However, unlike Earth, there are no oceans of water to soak up the majority of impacts that do make it through the atmosphere. They all hit the surface somewhere.

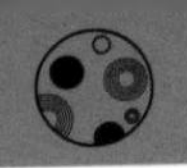

Mental Health

One of the biggest threats to Martian astronauts is internal, not external. The psychology of being cut off from home in an alien environment is a serious danger. Everyone and everything you love is stranded 50-400 million kilometers (30-240 million miles) away, barely visible as a tiny blue dot in the sky.

It will take homesick travellers at least seven months to get home. Even simple video communication is far from straightforward. Messages take 12.5 minutes to travel the average 225 million kilometers (139 million miles) back home and a reply will take a similar time to get to you. A conversation with your nearest and dearest would be very stop-start.

You'll also have to deal with being confined to a small space with the same people for prolonged periods, particularly on the journey there and back.

Between 2015 and 2016, six scientists spent a year isolated in a dome that was eleven metres wide and six metres high (36 x 20 foot), cut off from civilization in a remote part of Hawaii. A 20 minute communications delay with the outside world was imposed to simulate being on Mars. The participants had to learn to deal with conflict differently in the knowledge that they couldn't simply walk away from an argument.

Discovery & History

. . . there will be life on Mars, because we will bring it there.

Buzz Aldrin

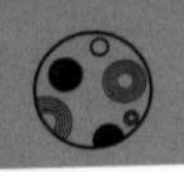

Formation

Like all the planets of the solar system, Mars wouldn't be here without the sun. Around five billion years ago an interstellar cloud of gas and dust buckled under its own weight and collapsed. Temperatures and pressures soared until a brand new star ignited – which we know today as the sun.

Some material was left over, spinning around the infant star in a flat, dark band called a protoplanetary disc. Over a period of 100,000 years, gravity started clumping the disc together to form large chunks called planetesimals. The inner solar system was too hot for water and many gases to survive, so these planetary building blocks were made of materials with high melting points: rock and metal. Many of them collided to fashion the rocky planets. The energy from these collisions kept the rock and metal molten, allowing gravity to round out Mars into a spherical shape.

The heaviest materials – iron and nickel – sank to the middle of the planet. The outer layers of Mars cooled into a crust, but the weight of these layers kept the core molten and drove early volcanic activity. The Red Planet has changed significantly since those early days: and exactly what has happened in the interim is the subject of much ongoing scientific scrutiny.

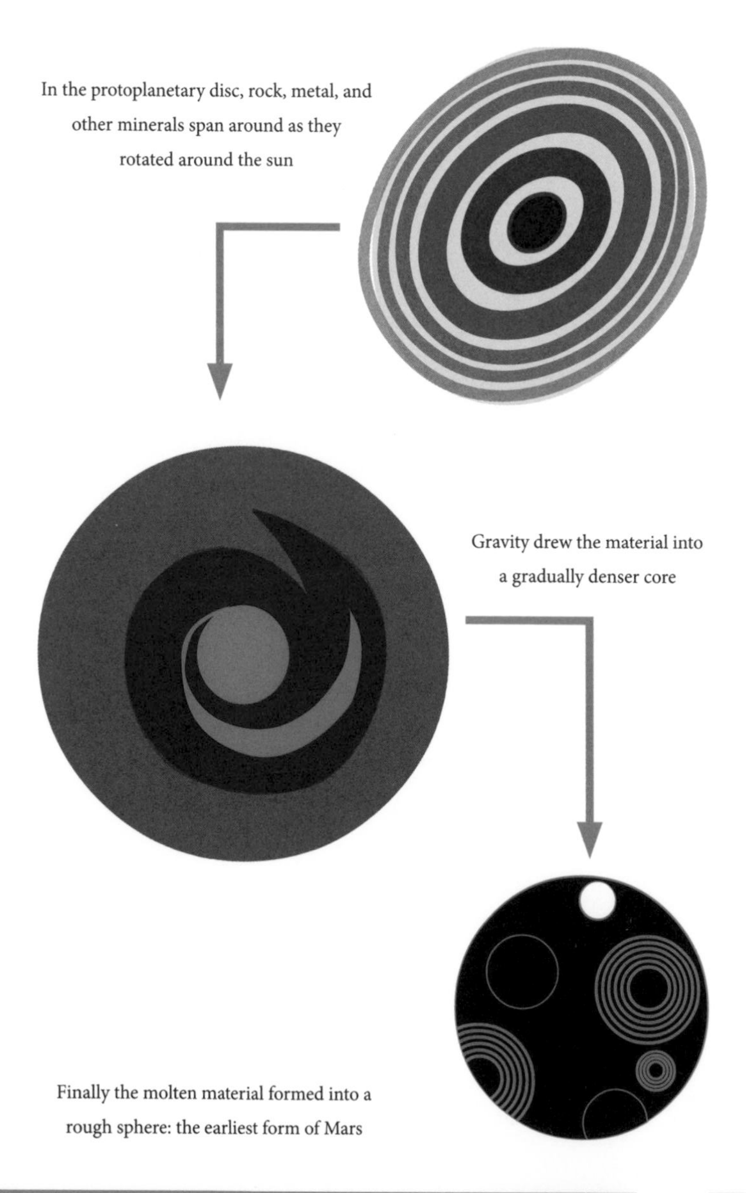
In the protoplanetary disc, rock, metal, and other minerals span around as they rotated around the sun
Gravity drew the material into a gradually denser core
Finally the molten material formed into a rough sphere: the earliest form of Mars

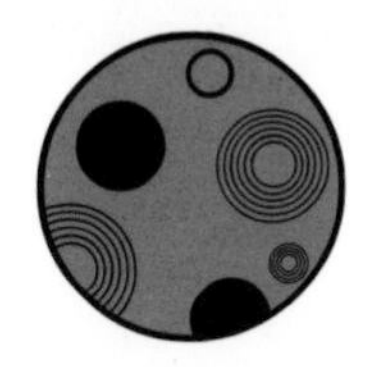

Ancient Observations

Humans have been tracking the passage of the Red Planet across the sky for thousands of years. Babylonian astronomers kept detailed records of its movement as early as 1500 BCE. They named it *Nergal* after the god of fire, war, and destruction.

A depiction of Mars appears on the ceiling of the tomb of the Ancient Egyptian pharaoh Seti I. In the 4th century BCE, the Greek philosopher Aristotle noticed one night that Mars disappeared behind the moon, and came to the correct conclusion that the planet must be further away.

At this point no-one knew that this orangey-brown light in the sky was another spherical, rocky world like the Earth. The word planet comes from the Greek *asteres planetai* which means wandering star. Skywatchers could observe five of these "planets" – stars that appeared to move while the other stars stayed fixed in their constellations.

Our modern names for them – Mercury, Venus, Mars, Jupiter, and Saturn – come from Roman gods. Like the Babylonians, the Romans named Mars after their god of war, probably because its colour resembles blood. It was known as the fire star in East Asian cultures, *Ma'adim* or *the one who blushes* in Hebrew, and the deity Mangala in Hindu religious texts.

Roman statue of Mars, the God of war

Mars in the Copernican Revolution

Watch Mars in the night sky over the course of several months and you'll notice something weird: it doesn't just wander in one direction. At first it appears to move through the constellations one way, only to stop and double back on itself.

Astronomers call this *retrograde motion*. Ancient skywatchers struggled to explain it because they believed the sun, the moon, and the planets all orbited the Earth. So they invented epicycles

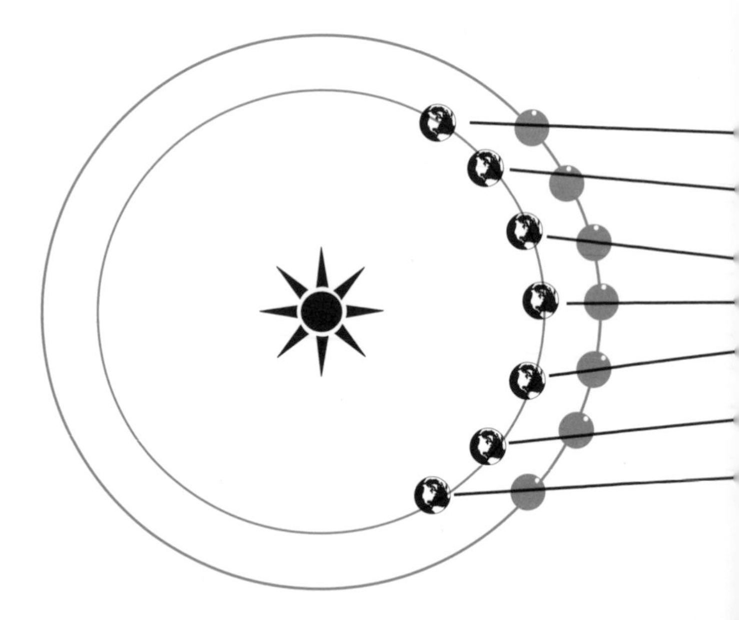

and deferents – essentially circles within circles in order to get their Earth-centered view to fit.

Then, in the 15th century, Polish mathematician Nicolaus Copernicus said he could explain the retrograde motion of Mars far more simply. All you had to do was relegate the Earth to the status of being just another planet orbiting a central Sun.

Under this model we see Mars seemingly change direction in our sky as we overtake it on our shorter journey around the sun (as illustrated in the diagram below).

This idea faced huge opposition from the Church, who were wedded to the biblical account of Earth at the centre of creation. However, faced with a mountain of evidence to the contrary, they eventually had to concede that Copernicus was correct.

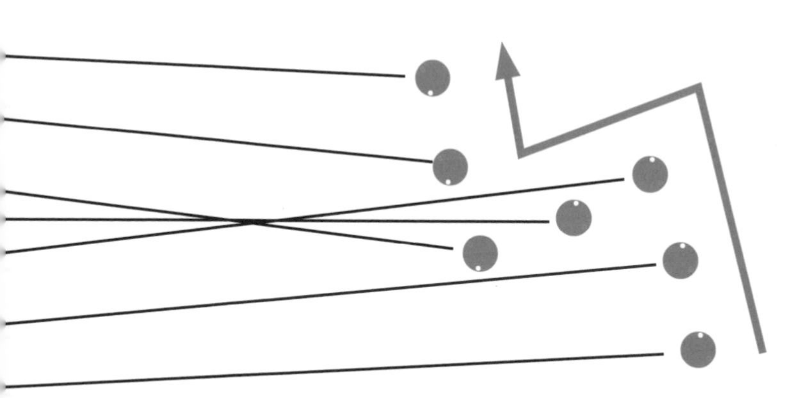

We regularly overtake Mars as we both orbit around the sun at different speeds. This results in us seeing Mars following a jagged route across the night sky.

Views with Early Telescopes

The Italian astronomer Galileo Galilei was probably the first person ever to see Mars through a telescope in 1610. He noted that the apparent size of the planet changed over time, which suggested that its distance from us was constantly shifting.

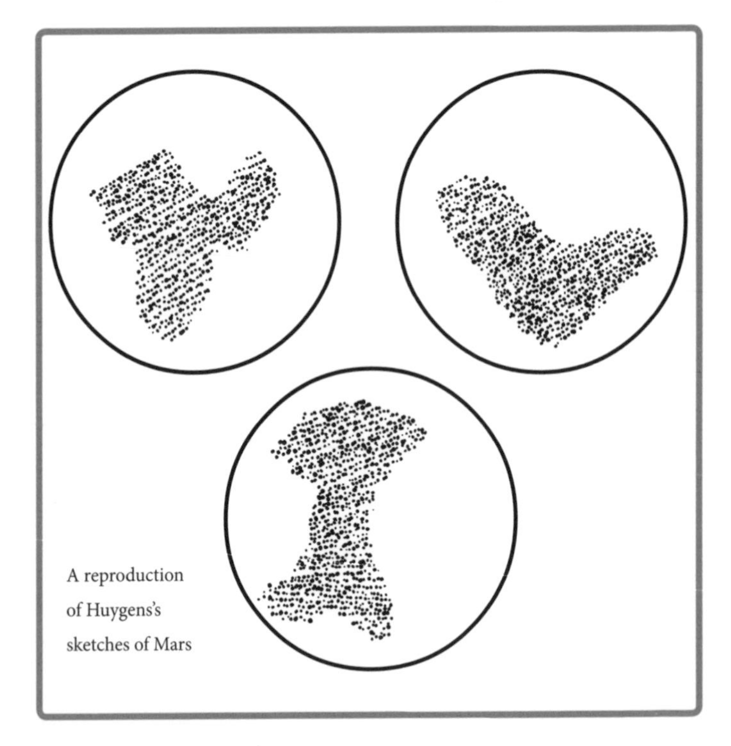

A reproduction of Huygens's sketches of Mars

An early image of a "Dutch telescope" from 1624. Galileo borrowed some of the principles of his telescope from a Dutch manufacturer

Dutch skywatcher Christian Huygens drew the first detailed maps of the Martian surface in the 1650s. His sketches clearly show the dark region that is now known as Syrtis Major and possibly one of the polar ice caps.

In the 1660s, Italian astronomer Giovanni Cassini used distinctive surface features to track Mars's rotation speed. Remarkably, his answer was within three minutes of today's accepted value. He was also able to estimate the distance to Mars to within 10%. In the late eighteenth century, the British astronomer William Herschel noted that the southern polar caps of Mars shrink and grow over time, suggesting the Red Planet has swinging seasons.

Then, in 1877, Mars and Earth moved within just 56 million kilometers of each other, gifting astronomers an unprecedented view of the planet. It was during this time that Asaph Hall discovered the moons of Mars and detailed maps showing an unsubstantiated system of "canals" began to appear (see page 52).

The Discovery of Phobos & Deimos

When American astronomer Asaph Hall finally tracked down the two tiny moons of Mars in 1877 it was a curious case of fact emulating fiction. In 1726, writer Jonathan Swift had included details of two Martian satellites in his famous book *Gulliver's Travels*. Reportedly inspired by this tale, Hall went looking for the moons for real using a telescope located in Foggy Bottom, Washington D.C.

But even then he almost didn't find them. After an exhausting search he was on the verge of giving up, telling his wife Angeline that he was thinking of jacking it in. She persuaded him to continue looking and, as a result, he finally spotted Deimos on 12 August 1877 and Phobos six days later. Their names mean "terror" and "fear" and are derived from the twin sons of the God of War who followed him into battle.

When the *Mariner 4* probe sidled up to the satellites in 1965 it snapped close-up images of the moons for the first time. A large crater was revealed on the surface of Phobos. A naming committee led by Carl Sagan decided it should be called Stickney – the maiden name of Hall's wife.

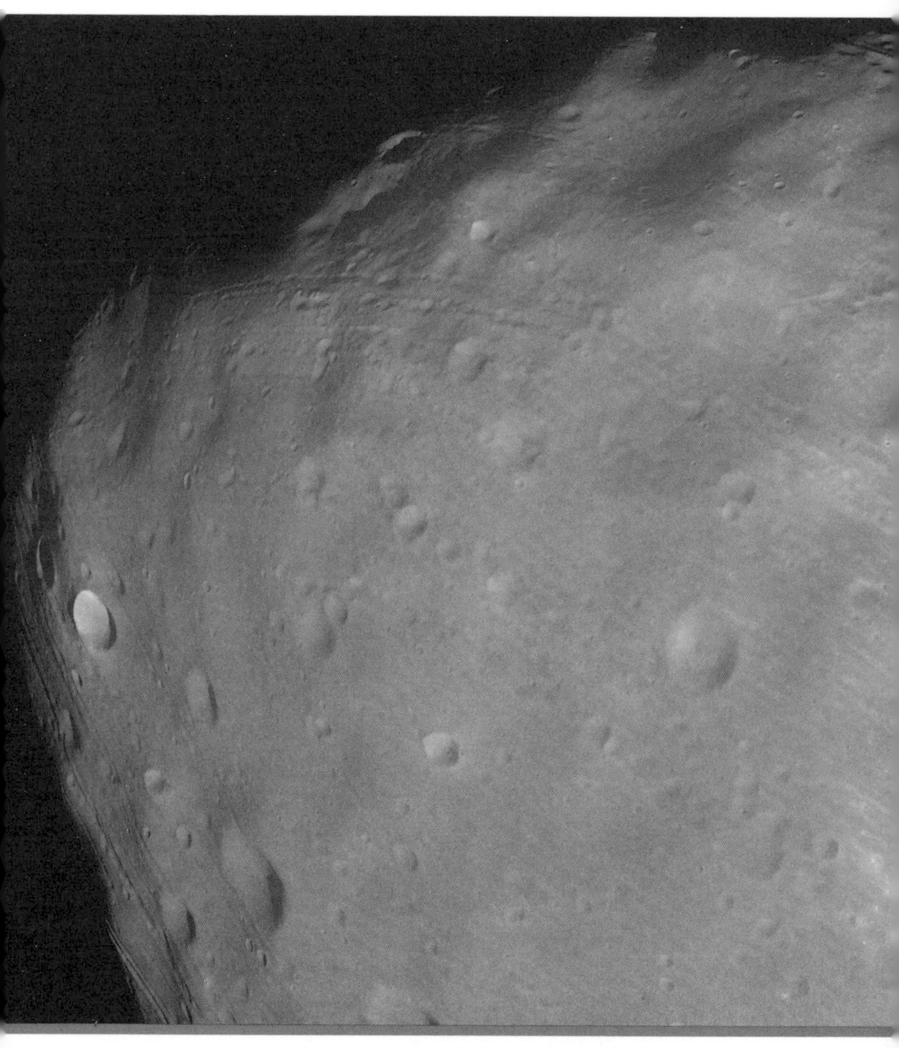

A dramatic view of Phobos taken from a distance of 6,800 km by NASA's *HiRISE* camera (pictured in the artist's rendition opposite)

Canals on Mars

Giovanni Schiaparelli, director of the Brera Observatory in Milan, began to draw maps of Mars during its close approach to Earth in 1877. He noticed a series of intersecting straight lines criss-crossing their way across the Martian surface. He called them "canali", the Italian word for channels.

Percival Lowell using the Lowell Observatory telescope in 1914

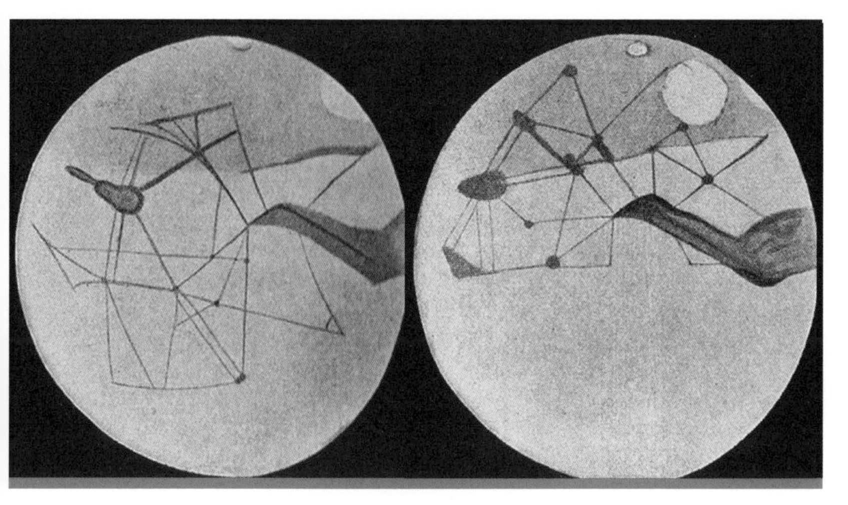

Lowell's drawings of the "canals"

However, this was mistranslated into English as canals and there is a big difference between channels and canals. Channels can be carved out naturally by running water whereas canals are artificial waterways constructed by living things.

All of a sudden, in the public consciousness Mars became a world inhabited by a race of master builders. Perhaps they'd engineered these waterways as a means of ferrying water from the icy poles to arid regions around the equator.

Prior to this, there had been a few science fiction stories about Martians. But within a few years there was an explosion in such tales. This trend intensified after American astronomer Percival Lowell (pictured opposite) published his own canal maps in 1895. H.G. Wells's *The War of The Worlds* was released just three years later in 1898. But it was, after all, just a mistranslation. Not only are there no canals on Mars, there aren't even the channels originally described by Schiaparelli. It was all an illusion.

Climate Change

Something catastrophic has happened to Mars. Today it is a bone-dry desert of a planet. Yet there is increasingly weighty evidence that it was a lot more plush billions of years ago. It may even have sported oceans a mile deep in places and covering at least a fifth of its surface. It's possible that at one point in its history it wasn't that different from the Earth.

What changed? Most astronomers point the finger at Mars's core. Originally the planet's centre was molten, much like the Earth's. Sloshing iron would have generated a magnetic field, a buffer against the ravages of the solar wind. However, as it is a smaller world, there isn't as much planet to crush down on the core. Eventually the iron cooled and solidified and the magnetic field vanished over a period of time.

Slowly but surely, the solar wind – a gale of charged particles blowing away from the sun – pecked away at the Martian atmosphere leaving just a thin wisp of carbon dioxide today. Most of the planet's stock of water was lost to space as temperatures and the atmospheric pressure dropped.

Just a sixth of the water remains, largely frozen in the planet's polar caps. There is much debate about whether the existence of its early oceans should lead us to believe that Mars was habitable and whether life ever got started there.

Image showing the most likely location of the ancient ocean at the top of the planet

Mariner 4 and *Mariner 9*

On 28 November 1967, NASA launched the *Mariner 4* probe (pictured below) to explore Mars. A little under nine months later it became the first human artefact to fly past the planet, and it returned the first close-up images of Mars.

The first digital image was hand-colored by scientists using pastels from an arts and crafts store to shade in individual pixels based on a numerical values on a printout – they were literally painting by numbers. They were too excited to wait for the mission

Mariner 4

A view of Olympus Mons taken from *Mariner 9*

computer to churn the data. A total of 21 grainy black and white images were taken, the rest being processed by the computer. They confirmed the presence of impact craters on the Martian surface. The first spacecraft to enter orbit around Mars – or any other planet for that matter – was NASA's *Mariner* 9 probe. It finally reached the Red Planet in November 1971. This time over 7000 images were returned to Earth, although only after a massive global dust storm had cleared and the surface became visible. The rover spotted the massive Olympus Mons volcano and astronomers named the vast canyon system stretching across the Martian equator "Valles Marineris" after the probe.

The *Viking* Landers

On 20 August 1975, NASA launched the *Viking 1* mission to Mars. *Viking 2* joined it en route to the Red Planet nearly three weeks later. Both reached Mars in the summer of 1976 and each spent a month scouting for safe places to land.

Trenches dug by the landing device on the surface of Mars

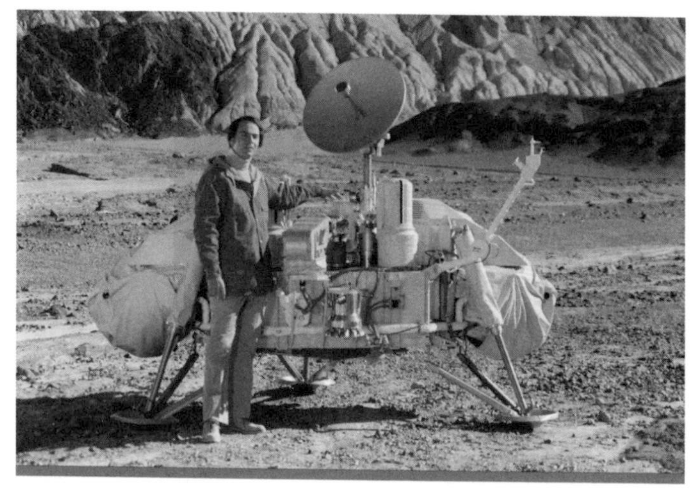

Carl Sagan with a model of a *Viking* lander

They both touched down in the planet's northern hemisphere: *Viking* 1 in Chryse Planitia on July 20 1976, and *Viking* 2 in Utopia Planitia on September 3.

As well as sending back stunning images of a vividly rich red surface, the landers had robotic arms which were used to gather up samples from their immediate environments.

Experiments on board baked, scratched, and sniffed the Martian soil looking for any hints of biology. Remarkably, one of the experiments came back positive, although scientists now believe this was a false positive caused by non-biological chemical reactions.

Viking 2 lasted almost four years before a battery failure ended its days. Its sister lander held out for more than six years before human error during a software update crippled its antenna. It held the record for the longest continuous operation on the Martian surface until it was surpassed by *Opportunity* in 2010.

Pathfinder and Sojourner

The *Viking* probes of the 1970s were static, restricted to examining regions their robotic arms could stretch to. To explore Mars in more detail you need a rover – a robot with wheels capable of moving from site to site in search of signs of water and life. Such machines had already trawled the lunar surface in this period, but none had travelled across the Red Planet.

Launched in 1996, NASA's *Pathfinder* mission carried the *Sojourner* rover to Mars. Like *Viking 1*, it landed in the Chryse Planitia

A rover from *Sojourner* drills into a rock.
From the same panoramic image, the section opposite shows the ramp down from the landing craft and the Twin Peaks mountains.

region and was dropped onto the surface inside an air bag that bounced a few times before coming to a standstill.

Only 65 cm (26 inches) long and trundling on six wheels, the rover sidled up to nearby rocks named Barnacle Bill, Yogi, and Scooby-Doo by mission scientists. Its maximum speed was just one centimetre per second (36 metres per hour). *Sojourner* was only designed to operate for seven Martian days, but held out for a total of 83. It covered a total of 100 metres in that time. Over 16,000 images were returned to Earth, including spectacular photos of the Martian sunset. Almost ten million scientific measurements were taken of Mars's winds and atmosphere.

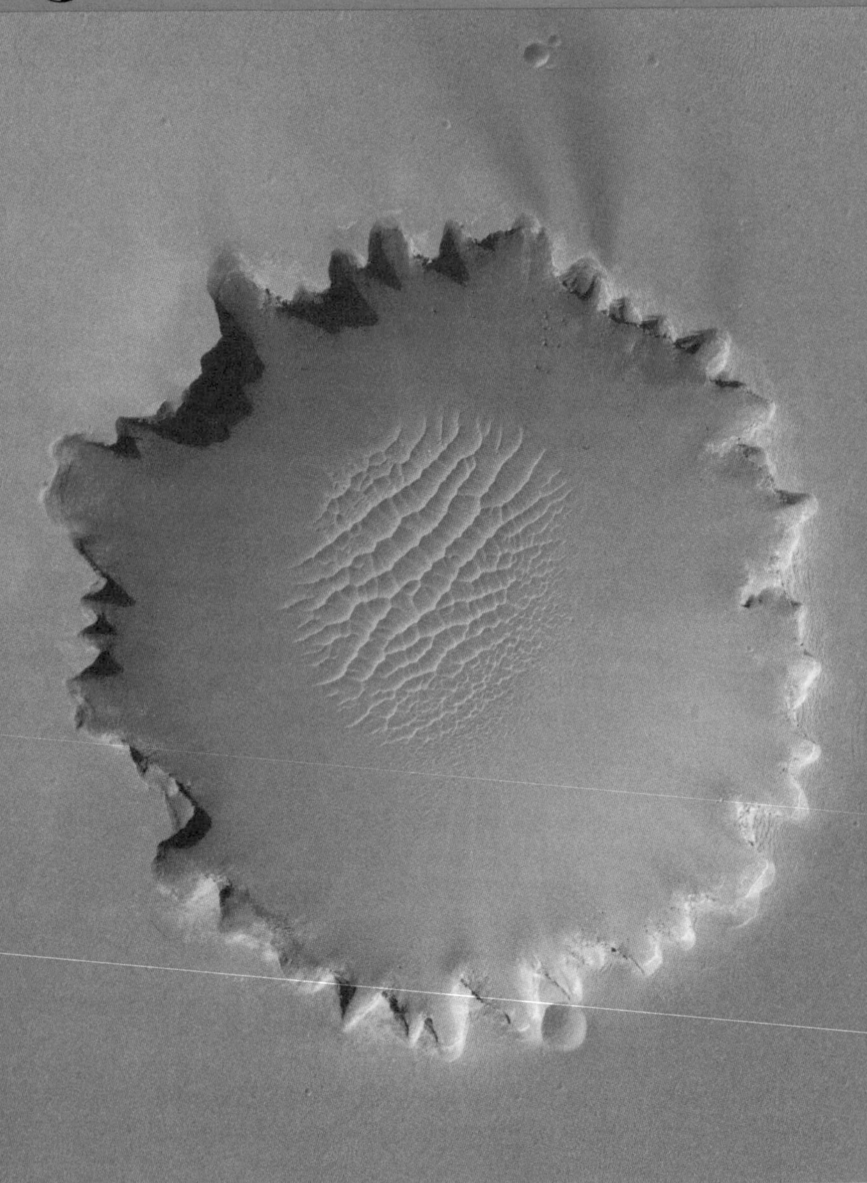

Spirit and *Opportunity*

Sojourner blazed a trail for the rovers that followed. In 2004, the twin *Spirit* and *Opportunity* rovers emerged from their inflatable cocoons to begin life on a new planet. *Spirit* touched down in Gusev crater, just south of the Martian equator, while its twin landed almost exactly on the equator in Meridiani Planum. Both rovers had six wheels like *Sojourner*, but were around 2.5 times longer. Originally designed to last just 90 Martian days, *Spirit* lasted until late 2009, when it became trapped in deep, sandy terrain.

Far exceeding expectations, some of the highlights of the overall mission included drilling into the Adirondack rock, finding evidence of past water, driving around Bonneville crater, and trundling around the Columbia Hills.

Remarkably, at the time of writing, *Opportunity* is still operational having survived a staggering 14 years on the surface, easily trumping the record of *Viking* 2 and covering a distance greater than the Olympic marathon.

Along the way it has survived the onslaught of several fierce dust storms and getting lodged deep in a sand dune from which it took six weeks to free itself. Within its first year of operation it had found an intact meteorite on the surface and it subsequently visited the Erebus, Victoria, and Endeavour craters.

The Victoria Crater photgraphed by *HiRISE*

Curiosity

The success of *Spirit* and *Opportunity* inspired scientists to become even more ambitious. In 2012, they landed the car-sized *Curiosity* rover on Mars. Far too big to be dispatched by inflatable airbag, it was lowered onto the surface by a futuristic looking sky crane. Its name came from 12-year-old Kansas schoolgirl Clara Ma who had suggested it via a public essay competition. *Curiosity* landed in the bed of Gale crater, in the shadow of the 5.5 kilometer-high (3.4 mile-high) Mount Sharp (Aeolis Mons). The treads on each of its six wheels are designed to leave a pattern in the Martian surface that spells out JPL in Morse Code (.– – – . – –. . – –. .). As well as a hat-tip to NASA's Jet Propulsion Laboratory where it was built, this helps the rover's cameras work out how far it's travelled. To mark the end of its first year of operation in 2013, the rover played Happy Birthday to itself by vibrating its soil sample analyser at the appropriate frequencies. It was the first time a song had been "played" on another planet. The mission has revealed that Gale crater was probably once a lake filled with rain running off its ancient slopes.

Local Produce

The first footfalls on Mars will mark a historic milestone, an enterprise that requires human tenacity matched with technology to anchor ourselves on another world.

Buzz Aldrin

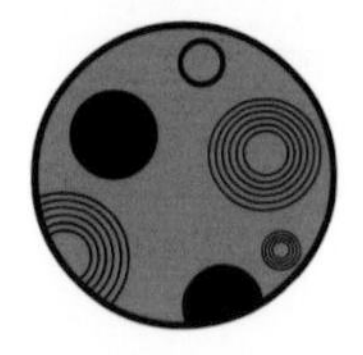

Rust

Mars is famous for its ruddy appearance, a distinctive hue provided by large quantities of iron oxide (rust) in its surface material. The Red Planet has more than twice as much surface rust as the Earth despite being smaller.

All the rocky worlds of the solar system have iron in them – astronomers believe that the planets emerged from a spinning cloud laced with it. Most of Earth's iron sank towards the middle of the planet where it created a chunky iron core.

However, on Mars – a planet with weaker gravity – a higher proportion stayed on the surface. Much of the surface iron rusted during the planet's warmer and wetter youth.

Then, as the planet dried out over billions of years, fierce winds eroded its rocks to create the fine red powder that coats the surface of Mars. This layer is only two metres deep at most.

A bed of dark, volcanic rock called basalt lies underneath. So, without its thin dusting of rust, Mars would be much closer to a charcoal grey and wouldn't have been named after the Roman god of war or be as easy to spot in the night sky.

Artist's impression of a rover designed for the *Mars 2020* mission. Based in many respects on previous rovers such as *Curiosity*, such future rovers will have drills to explore deep below the Martian rust

Methane

One of the biggest mysteries of Mars is the persistent presence of the gas methane in its atmosphere.

Methane is CH_4 – one atom of carbon stuck to four atoms of hydrogen. The high amount of ultraviolet light from the sun striking Mars supplies enough energy to prise these atoms apart, destroying any methane in the process. The maximum lifespan of any one methane molecule is 600 years. So if we're observing sizeable quantities of Martian methane today then it must have entered the atmosphere relatively recently.

Where did it come from? Well, 90% of Earth's methane is produced by biology. If the same is true on Mars then some ancient, long-dead lifeforms probably aren't responsible – that methane would have been destroyed by now. If the gas is indeed biological in origin, there might have been living things on Mars an astronomical heartbeat ago. Maybe they're there even now.

But before you get too excited about sharing your vacation on Mars with bona fide alien life forms, there are other possibilities. Maybe the methane was created by ancient, extinct life but was trapped in the Martian ice only to slowly escape over time. Or maybe old volcanoes are to blame and the gas is similarly only now emerging from underground.

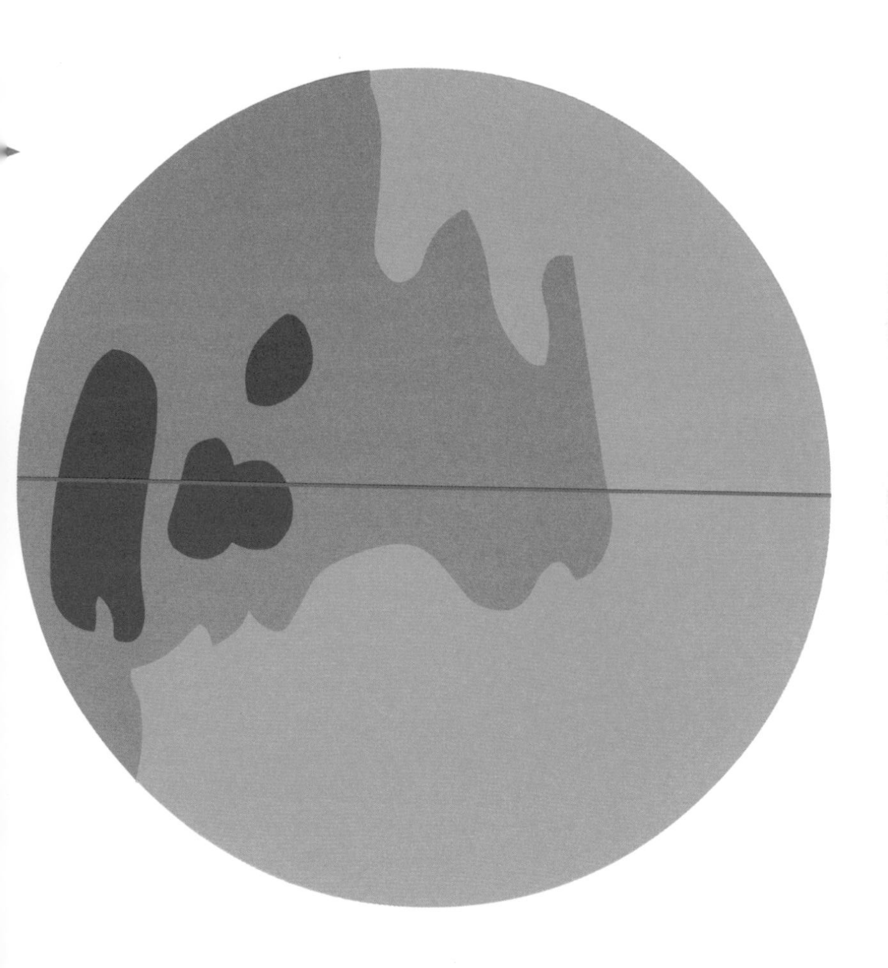

Methane concentrations are higher (darker shades in the image) in equatorial areas of the planet

Dry Ice

Mars has northern and southern polar caps just like Earth. However, temperatures can plummet so low in Martian winters that trillions of tonnes of its thin, carbon dioxide atmosphere also get frozen onto the ice caps in the form of dry ice. That's the same substance that is used at music concerts to create a smoky effect. Observations from orbiting spacecraft suggest 15-30% of the planet's CO_2 regularly undergoes this change.

The northern layer of dry ice is only a metre or two thick and disappears once winter is over – increased temperatures return the carbon dioxide to the Martian atmosphere.

However, due to the planet's unusual seasonal variation, the southern dry ice layer is much thicker (8 metres or 26 feet) and never completely vanishes, even during its summer period.

NASA's *Mars Reconnaissance Orbiter* has even spotted carbon dioxide snow clouds around the southern polar cap during winter. Measurements suggest particles within the cloud are big enough to fall as dry ice snowflakes.

It is thought to be the only place in the solar system where it snows carbon dioxide and it shows that dry ice is deposited onto the ice cap both as falling snow and as ground level frost.

Places to Visit

The gladdest moment in human life . . . is a departure into unknown lands.

Sir Richard Francis Burton

The Polar Caps

The ice caps of Mars are in a state of constant flux as temperatures vary throughout the seasons. When dry ice turns back into a gas, violent winds blow off the poles at speeds of up to 400 kilometers (250 miles) per hour. These gusts sculpt nearby loose sand into a series of spectacular dunes. From orbit, parts of the dunes look like grass or seaweed, but these dark patterns are just avalanches in the sand.

Perspective view of the Chasma Boreale

Dark sand cascades resembling trees are visible
in the upper left hand area of this image

The northern pole is dominated by the vast Chasma Boreale – a 560 kilometer long, 2 kilometer deep (350 x 1.2 miles) canyon that nearly cuts the cap in half. Astronomers believe it predates the ice cap and it grows with each passing winter. The area around the southern cap is covered in unusual looking features not seen on Earth including araneiforms – dark, spider-like projections.

When the south pole emerges from its deep winter, intensfying sunlight penetrates the ice and creates pockets of gas underneath it. As the gas bursts through vents in the ice, it carves out channels all emanating from the same point. Sublimation – carbon dioxide turning directly from a solid to a gas – is also responsible for turning areas of Mars's southern polar region into terrain that resembles Swiss cheese.

The Tharsis Bulge

Mars boasts a vast, volcanic plateau stretching 5,000 kilometers (3,100 miles) from Amazonis Planitia to Chryse Planitia. The size of a continent, this Tharsis region covers 25% of Mars and is home to three towering volcanoes, each of which is higher than Mount Everest: Arsia Mons, Pavonis Mons, and Ascraeus Mons. Together they are known as the Tharsis Montes and were first spotted by the *Mariner* 9 space probe in 1971.

The highest volcano on the planet – Olympus Mons – sits just to the west of the region, while Valles Marineris encroaches on its eastern slopes. The bulge, which spans the equator in its western hemisphere, was formed by lava pooling under the Martian surface before bursting through around 3.7 billion years ago. The latest observations suggest this lava gradually formed the Tharsis Montes one by one, starting with Arsia Mons, which is at the southern end.

This enormous upwelling of a billion billion tonnes of material from inside the planet twisted and contorted the Martian terrain, causing Mars to tilt over by at least twenty degrees.

For this reason some researchers point to the formation of the Tharsis bulge as one of the major reasons why the Red Planet's climate took a dramatic downturn.

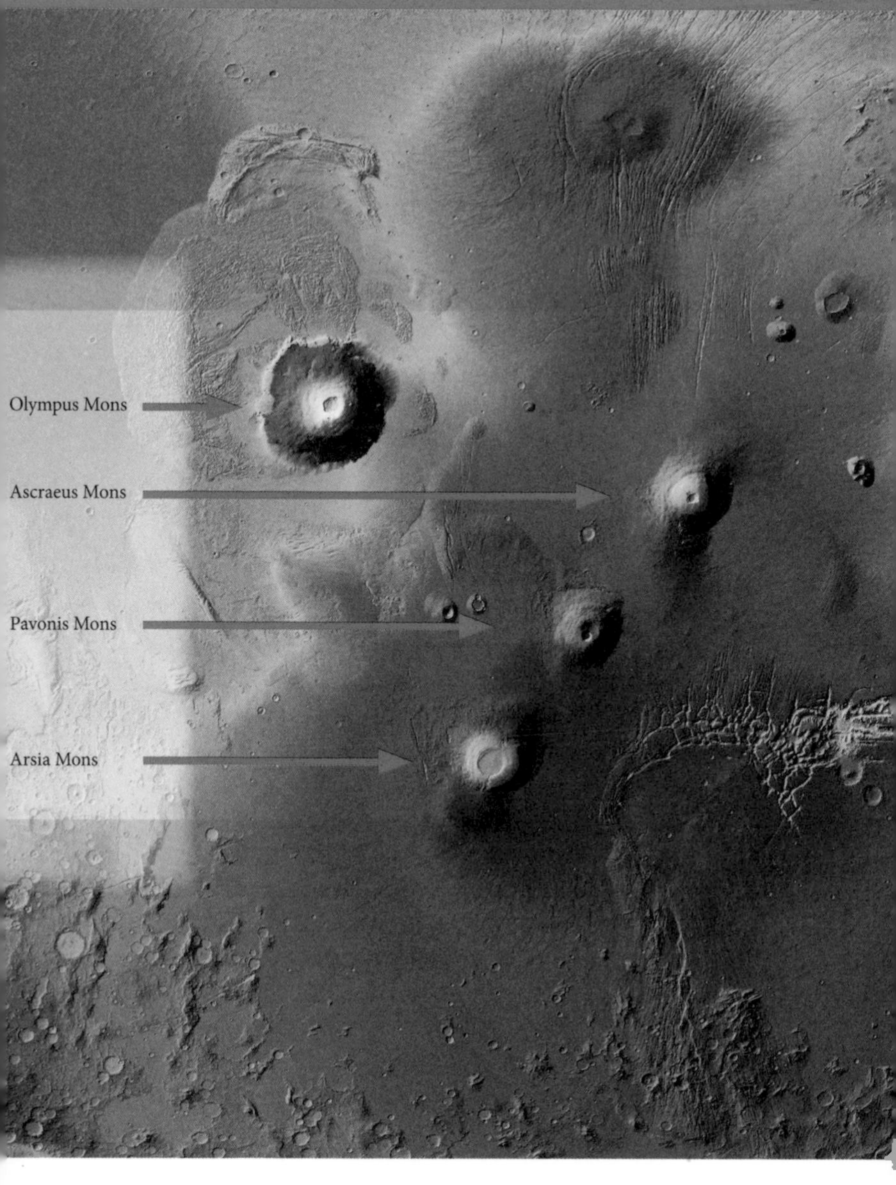
Olympus Mons
Ascraeus Mons
Pavonis Mons
Arsia Mons

Gale Crater and Mount Sharp

The 154 kilometer-wide (96 mile-wide) Gale crater was the landing site for NASA's *Curiosity* rover in 2012. It was selected due to the presence of clay and sulfate minerals there – both tend to be deposited by water. Scientists now believe that the crater, which formed from a giant impact around 3.5 billion years ago, was once a vast lake. Chemical analysis indicates the water would have been clean enough for humans to drink.

Carbon, hydrogen, nitrogen, oxygen, phosphorus, and sulfur – all key ingredients for life – have been discovered in the crater, leading to speculation that it was once home to microbes. That life was either wiped out once the water vanished, or it is just possible that it is still clinging on in underground aquifers. A series of run-off deltas and fans indicate that the amount of water in the lake regularly changed over millions of years. These features include the Pancake Delta and the Peace Vallis Fan.

Ancient rivers emptying into the lake deposited sediment in the centre of the crater. After the planet's climate changed and the water dried up, the planet's notorious winds sculpted this sediment on the lakebed into the towering 5.5 kilometer (3.4 mile) central peak known as Mount Sharp (officially named Aeolis Mons).

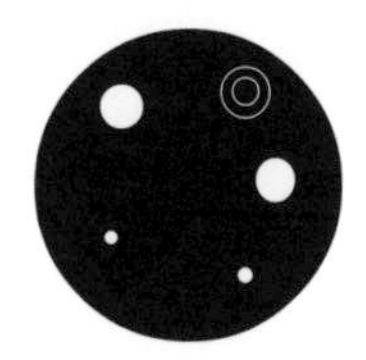

Terra Sirenum

In contrast to the smooth terrain of the Tharsis Bulge, the 4,000 kilometer-wide (2,485 mile-wide) Terra Sirenum (meaning "Land of Sirens") is a heavily cratered region to the south-west of Tharsis. The biggest of the impact scars is Newton crater at 300 kilometers (186 miles).

Seasonal flows appear on the crater rim in Martian spring and summer: they are possibly caused by the movement of salty water. A plethora of dry gullies show that larger volumes of water moved through the area during the planet's more temperate past. Additional evidence for historical water is provided by the chloride deposits spotted by the *Mars Odyssey* orbiter. Planetary scientists believe they were left behind when the water vanished. New gullies are still appearing today.

In 2013, NASA's *Mars Reconnaissance Orbiter* spotted a new channel on a nearby slope. This time the explanation was the action of carbon dioxide frost rather than water. Some of the local craters contain ice that is thought to have been deposited as snowfall in the past. Parts of the region show evidence of past tectonic activity and the ancient magnetic field of the planet. Other geographical features similar to those on Earth have also been found in the Terra Sirenum, including tongue-shaped glaciers, oxbow lakes, and lava flows.

Aerial view of the Terra Sirenum

Borealis Basin

Even a quick look at a map of Mars reveals a split personality. Its southern end is dominated by mighty highlands, yet the top is incredibly flat. The huge Borealis Basin (Northern Polar Basin) covers 40% of the planet.

This global dichotomy is the subject of much debate among planetary scientists. Some argue that it must be the result of an almighty ancient impact with a body at least 1,600 kilometers (1,000 miles) across – larger than Pluto. If it really is an impact basin, it would be the largest one in the solar system.

Such an event could also have spawned the moons Phobos and Deimos – debris thrown into orbit by the collision could have later coalesced into the two satellites. This contrasts with the traditional theory that they are asteroids that were captured from the nearby asteroid belt.

Advocates of an ancient impact point to magnetic anomalies in the southern half of Mars. The impact would have sent a massive shockwave careering through the planet, disrupting its crust on the other side. Whatever its origin, the low-lying basin was home to Mars's largest ocean before most of its water was lost to space or frozen into the terrain.

On Mars we can see five giant impacts, including the giant Borealis basin (top of globe), Hellas (bottom right), and Argyre (bottom left). The darker areas of the globe show the lowest altitudes.

Syrtis Major

On the border between the southern highlands and northern lowlands of Mars sits the distinctive dark feature of Syrtis Major. We've been looking at the area for a very long time – it was used by Dutch astronomer Christian Huygens in the 17th century to estimate the length of the Martian day. It was the first permanent feature seen on another planet.

Syrtis Major has been known by various names over the years including the Hourglass Sea and the Kaiser Sea. Its current name comes from maps of Mars drawn by Giovanni Schiaparelli in 1877.

It was originally classified as a low-lying plain and given the name Syrtis Major Planitia. However, we now know that it's an ancient, shallow shield volcano and so the name has changed to Syrtis Major Planum. (Planum is the official term for a plateau or high plain.) The western end is 4 kilometers (2.5 miles) higher than the eastern end.

The area's trademark darkness comes both from the presence of basaltic rock – formed from cooled lava – and a general lack of the planet's ubiquitous red dust. Its appearance changes with the seasons as the Martian winds redistribute what little dust there is into bright streaks downwind of its craters.

Viking Orbiter image of the Syrtis Major region, taken from 2,000 km (1,250 miles). The dark area to the right is Syrtis Major Planum. The large crater towards the bottom centre of the image is Huygens crater.

Hesperia Planum

Another region named by Giovanni Schiaparelli in 1877, Hesperia Planum is famous for its wrinkle ridges – tectonic structures formed when ancient lava flooded the plains here.

Hesperia means "lands to the west" a name which is derived from its position in the planet's western hemisphere. The western end of the plain is home to the Tyrrhenus Mons volcano. At only 1.5 kilometers (0.95 miles) high, it reaches a significantly lower altitude than the majority of Martian volcanoes.

This is probably because, rather than spewing lava onto the surface, the volcano erupted ash and dust that collapsed into shallower slopes. These diminutive peaks are known as paterae – the Latin for dishes. Tyrrhenus Mons shows significant signs of erosion, suggesting it is quite old.

The presence of several large craters nearby, formed when there were still large impactors flying around the solar system, further indicates this part of the surface is far from new.

One of these craters – a large ellipse 24.4 km (15.2 miles) long and 11.2 km (7 miles) wide – is shaped like a butterfly. The region gives its name to the Hesperian Period, one of the four geological ages of Mars (occurring after the Pre-Noachian and Noachian Periods, but before the present Amazonian Period).

A close-up image of Hesperia Planum

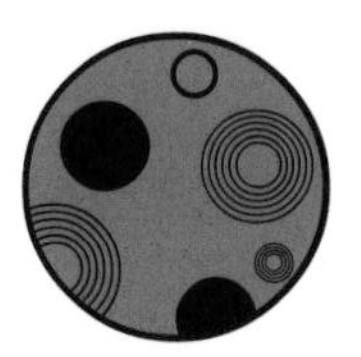

The Meridiani Salt Flats

Some of Earth's most spectacular geography can be seen in the vast, dry salt flats located in places such as the United States of America, Bolivia, and Chile.

The same is true of Mars. The Meridiani Planum, the region where the *Opportunity* rover landed in 2004, is also home to almost 50 square kilometers (19.3 square miles) of spectacular salt flats which have been compared to the Bonneville Salt Flats in Utah, USA.

As on Earth, these regions were once home to salt water lakes and when the water disappeared the salt was left behind. Estimates based on the thickness of the salt suggest that the Meridiani lake would only have been 8% as salty as Earth's oceans. So its water would have been reasonably fresh and maybe even habitable for simple microbial life.

More than 640 similar chloride deposits have been found across Mars, most notably in the Terra Sirenum. However, analysis of the Merdiani salt flats has pointed to the original lake still having been there approximately 3.6 billion years ago, around a billion years after the planet formed. That means it was one of the last vestiges of substantial liquid water on the Red Planet before its climatic downturn.

Volcanic ash deposit in the Meridiani Planum

The Water Flows of Garni Crater

Ever since the fiasco of the Martian canals confusion, humans have been captivated by the idea of life on Mars. And from what we know about biology on Earth, all living things need at least some amount of liquid water, irrespective of whether they are a tiny bacterium or a gargantuan blue whale.

So liquid water on the surface of Mars is always going to make astronomers sit up and take note. The trouble is that liquid water on Mars is a real rarity. The planet's puny atmospheric press means

that water normally skips out the liquid state altogether, jumping from ice to water vapour in a process known as sublimation. However, something different appears to be happening on the sloped walls of Garni crater. Here temporary dark streaks hundreds of metres long are seen etched into the landscape.

Analysis from orbit points the finger at the movement of liquid water as the culprit. It is thought this water is particularly salty and the salt helps lower the freezing point of the water, just as salt on our roads and walkways prevents ice from forming. It's the strongest evidence yet that Mars can play host to liquid water, albeit for short periods.

Orcus Patera

Deriving its name from its whale-like shape, this unusual depression in the Martian surface has stumped astronomers for decades. It was first imaged by *Mariner 4* in 1965, and has been studied extensively since by missions including *Mars Global Surveyor* and *Mars Express*. We still don't know how it formed.

Located near the Martian equator, halfway between the volcanoes Olympus Mons and Elysium Mons, it is 380 kilometers long and 140 kilometers wide (236 x 86 miles). At first it looks like an impact crater. However, the explosion of an impactor on contact normally blasts a circular depression in a planet's surface. If Orcus Patera really was formed by a stray asteroid, it would have come in at a very shallow angle.

Other scientists have pointed the finger at volcanism, noting its presence close to other clear signs of volcanic activity. It could also have been formed by tectonic activity – the shifting of the planet's surface over time.

The edge of the patera is dotted with criss-crossing, valley-like structures called graben. However, it seems they were chiselled out after Orcus Patera had formed. So scientists are still scratching their heads and that head-scratching may continue until field geologists can visit the area in person.

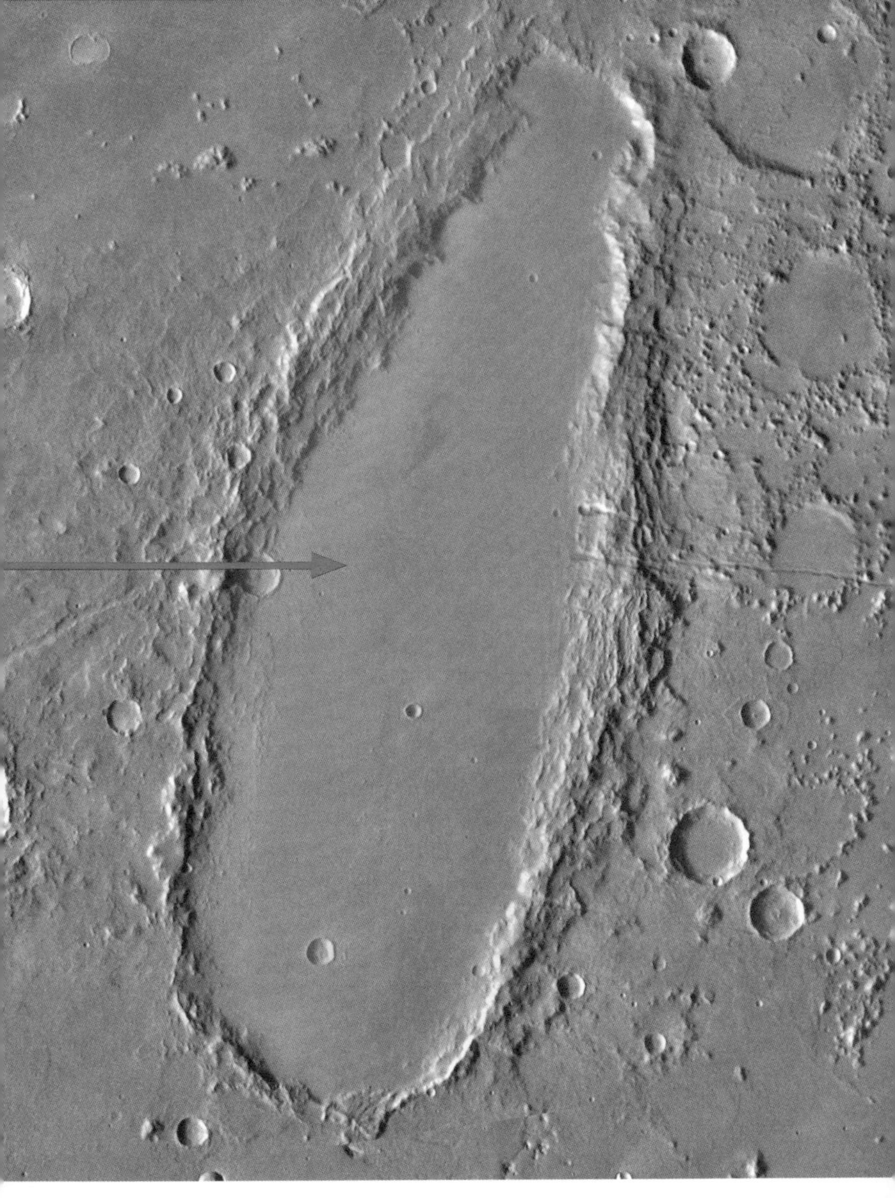

Olympus Mons

The extinct shield volcano Olympus Mons – or Mount Olympus – towers nearly 22 kilometers (13.6 miles) above the dry Martian surface (and 26.4 kilometers or 16.4 miles above sea level). That's around 2.5 times higher than the top of Mount Everest here on Earth. And it is only a fraction below the solar system's tallest summit – the central peak of the Rheasilvia crater, which is on the asteroid Vesta.

Yet ascending it isn't as difficult as you might imagine – the slopes are incredibly shallow with an average inclination of just five degrees. It won't even feel like you're climbing at all. Although don't expect to be able to see the peak when you start your climb – the volcano is so wide that the summit will be obscured beyond the horizon! It will take you weeks to complete the journey on foot. Those with shorter timescales should consider one of the many overnight sleeper shuttle trips on offer.

Just like Mount Everest, there are different routes up the mountain with varying degrees of difficulty. The slopes are shallower on the north-western flank of the volcano, but steeper when you take the south-eastern routes.

Your reward for reaching the top will be a spectacular view of not only the Martian landscape below, but also of the mighty caldera at the top. At over 60 kilometers wide and 3 kilometers deep (37 x 1.8 miles) those staying on the mountain can partake in a spectacular array of climbing and abseiling activities.

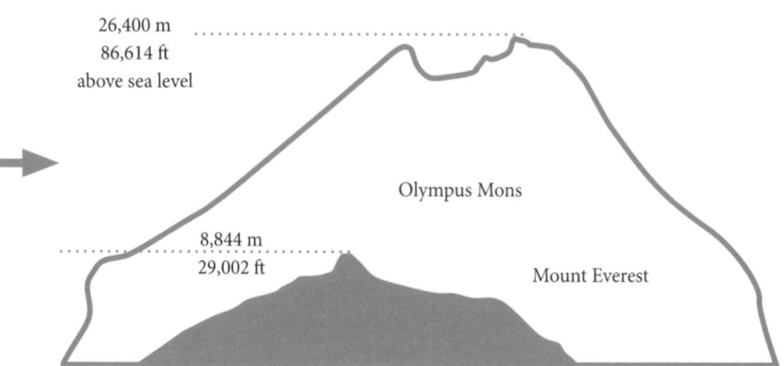
26,400 m
86,614 ft
above sea level
Olympus Mons
8,844 m
29,002 ft
Mount Everest

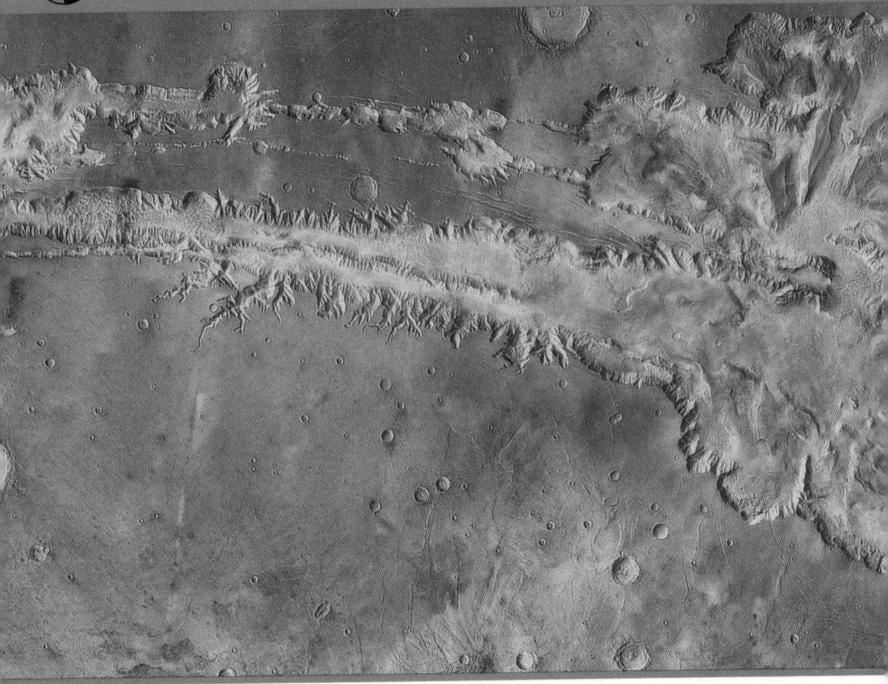

Valles Marineris

Along with Olympus Mons, Valles Marineris ("the Mariner Valleys") are the poster-children of Mars. Named after the *Mariner 9* mission that first photographed them, expect them to be on all the postcards and to dominate the gift shop. A huge canyon system with a total length of over 4,000 kilometers (2,500 miles), they stretch almost a quarter of the way round the Martian equator. That's as long as the United States is across. Exploring the entire area will take you many months. Unlike the Grand Canyon in Arizona, Valles Marineris weren't formed by an

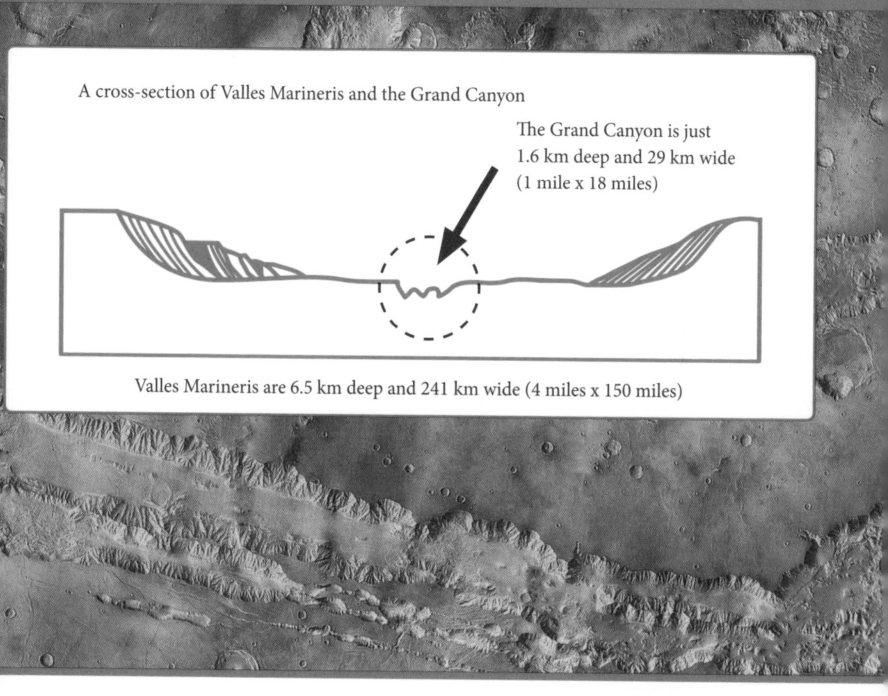

ancient river. Instead, planetary scientists believe that they're fissures in the Martian surface that cracked open when the Tharsis bulge formed. Erosion and landslides have since made them even wider: they have become more than 200 kilometers (125 miles) wide and 7 kilometers (4.5 miles) deep in places.

You will be crossing the jumbled terrain of Noctis Labyrinthus ("Labyrinth of The Night") on the journey from your accommodation at Pavonis Mons to the valleys. It is directly connected to the hotel's lava tubes and that's where your expedition will begin. As you venture further down the valleys you'll encounter a series of steep-sided canyons known as chasmata before reaching Chryse Planitia at the other end.

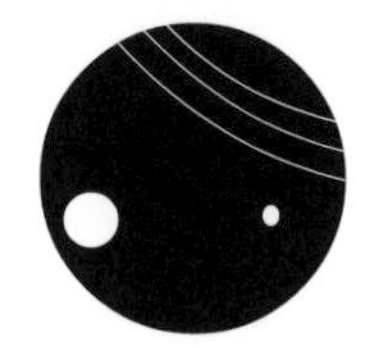

Hellas Impact Basin

Located approximately in the middle of the southern hemisphere of Mars, the Hellas impact basin is one of the largest in the solar system. At a whopping 2,300 kilometers (1,430 miles) wide, it is Mars's most obvious impact scar. It's around 8 kilometers (5 miles) deep, meaning that Mount Everest (8848m or 29,002 feet) would only just peek out over the basin's rim. It contains the lowest lying location on the planet, leading to an atmospheric pressure on its floor which is twice the Martian average.

Astronomers believe it formed when the Red Planet was blindsided by a huge impactor during a sustained attack 3.9 billion years ago known as the Late Heavy Bombardment. The same period is responsible for much of the large cratering on the moon.

Named after the Greek word for Greece by Giovanni Schiaparelli, there are many geological treasures to wonder at here. The Hellespontus Montes mountain range spans its western border and you'll find the Dao Vallis and Reull Vallis gullies on its eastern flank. The latter are thought to be outflow channels from back when the basin was filled with liquid water. The hills of Alpheus Colles undulate on the basin's floor and geologists have even spotted evidence of Martian glaciers in the area.

This angle gives a clear view of the Hellas Basin

Stay the Night on Phobos

This is one to take advantage of before you land on Mars. The Red Planet's gravity means day trips to Phobos from the Martian surface are prohibitively expensive. Stay the night there before tackling the white-knuckle ride down to the planet itself.

The most dramatic aspect of Mars's largest moon is the extremely low gravity – the gravitational pull there is just 0.05% of Earth's (and only 0.15% of Mars's). This means you need to be tethered to the Phobian surface at all times. Something that is as

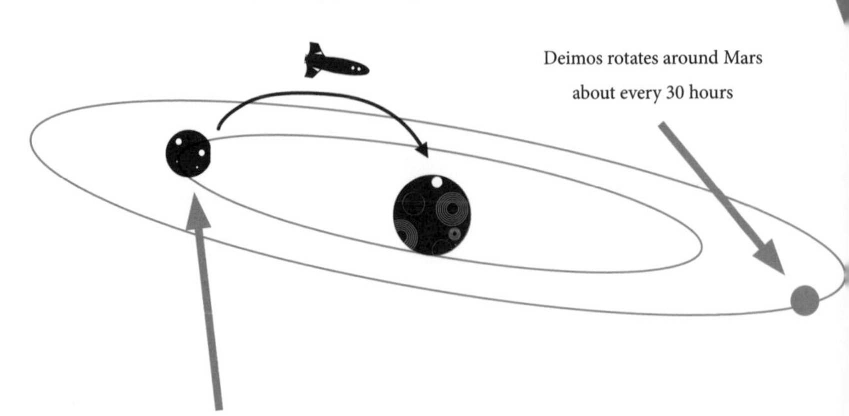

seemingly innocuous as a single oventhusiastic jump could all too easily see you float off into Martian orbit.

The satellite's biggest tourist attraction is a 90 metre (295 feet) straight-edged monolith surrounded by a wide, desolate expansive terrain. While some conspiracy theorists have suggested that this is evidence that intelligent species have been there before us, there is no indication that it is anything other than geological in origin. It could well be a boulder that was splintered off by an ancient impact.

Make the most of Phobos – the moon is gradually spiralling inwards towards Mars and eventually the planet's gravity will rip it apart into fragments, which will give Mars a temporary ring before its pieces fall down onto the planet.

A Trip to the Oxygen Factories

We humans take a lot of looking after and a major requirement is a plentiful supply of oxygen. On Earth we breathe in around eight litres (488 cubic inches) of air every single minute, even when we are resting. Just over 20% of that air is oxygen: it also contains nearly 80% of nitrogen and smaller amounts of carbon dioxide and other gases.

To survive on Mars we have had to build giant oxygen factories near the poles to generate all the O_2 we need. The staff open them up for public tours on the last Saturday of every month, but be sure to book in advance as spaces are limited and they prove very popular.

Our Martian oxygen comes from two separate sources: plants and water. Vegetation inside the huge, hydroponic food farms takes in carbon dioxide and pumps out oxygen. Some of this is siphoned off and stored for us to breathe.

However, that alone is not enough. At the oxygen factories, massive banks of solar panels also generate electricity which is passed through water from the ice caps.

This separates the water (H_2O) into hydrogen and oxygen. The hydrogen is stored for use as fuel and the oxygen is put inside pressurized containers and ferried to the various Martian habitats to maintain a breathable environment.

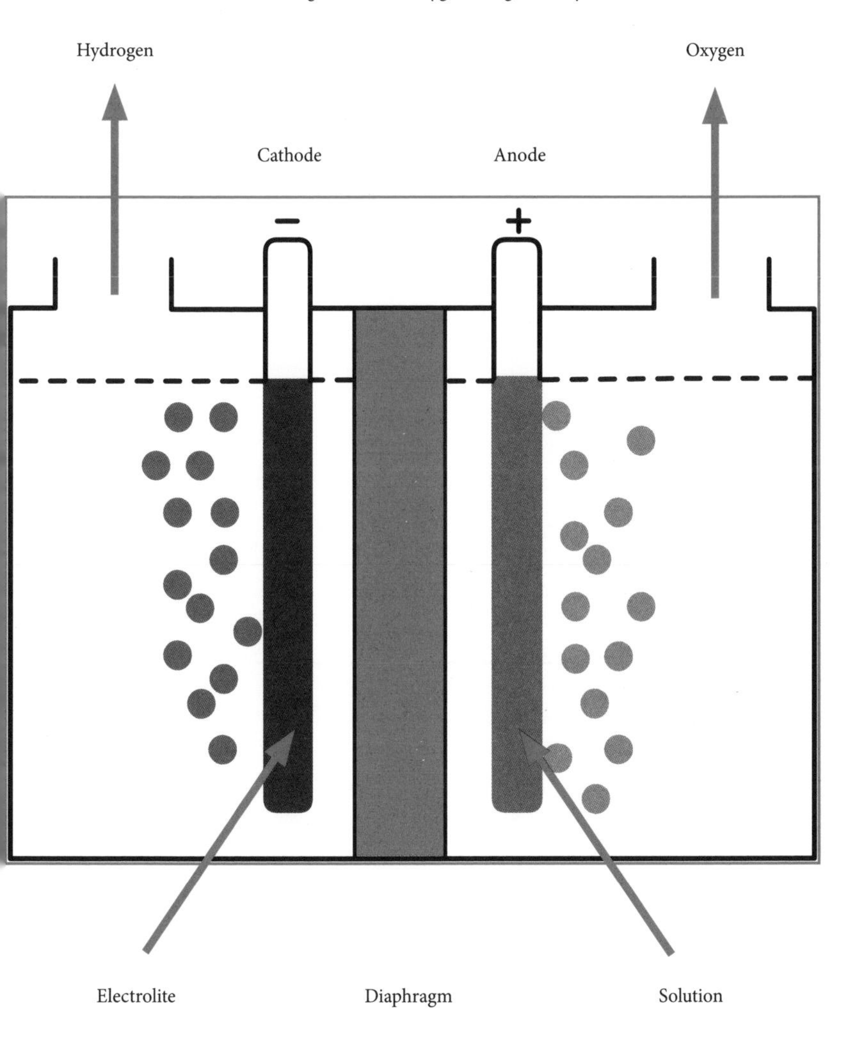
Water being turned into oxygen using electrolysis
Hydrogen
Oxygen
Cathode
Anode
−
+
Electrolite
Diaphragm
Solution

Visit the Food Farms

All Martians are vegetarian by necessity – there simply aren't the resources to sustain livestock as well as people. All food is grown inside specially constructed greenhouse domes that allow the environment to be carefully controlled.

Electric lighting is too expensive, so the chambers maximize the power of natural sunlight. The pressure inside is sufficient to walk around without a spacesuit and tours of the facility can be booked by prior arrangement.

Plants need soil, nutrients, water, and carbon dioxide to survive. The planet's huge ice supplies provide the water and the planet's atmosphere is mostly carbon dioxide, so there isn't a problem there either. Thankfully, the Martian soil is relatively friendly, naturally containing many of the nutrients required for healthy crops. It lacks enough potassium, so this is added. Toxic chemicals called perchlorates are also carefully removed prior to sowing. Crops are selected that provide the greatest amount of energy for the smallest amount of resources. This includes potatoes, wheat, soybean, radishes, peas, and leeks.

Experimental facilities are trialling the cultivation of insects on the Red Planet, a great source of nutrients. Crickets, for example, contain almost as much protein as beef. At present these areas are not open for public visits.

On the International Space Station, Shane Kimbrough harvests lettuce from the Veggie experiment, using the same technology as the greenhouses on Mars

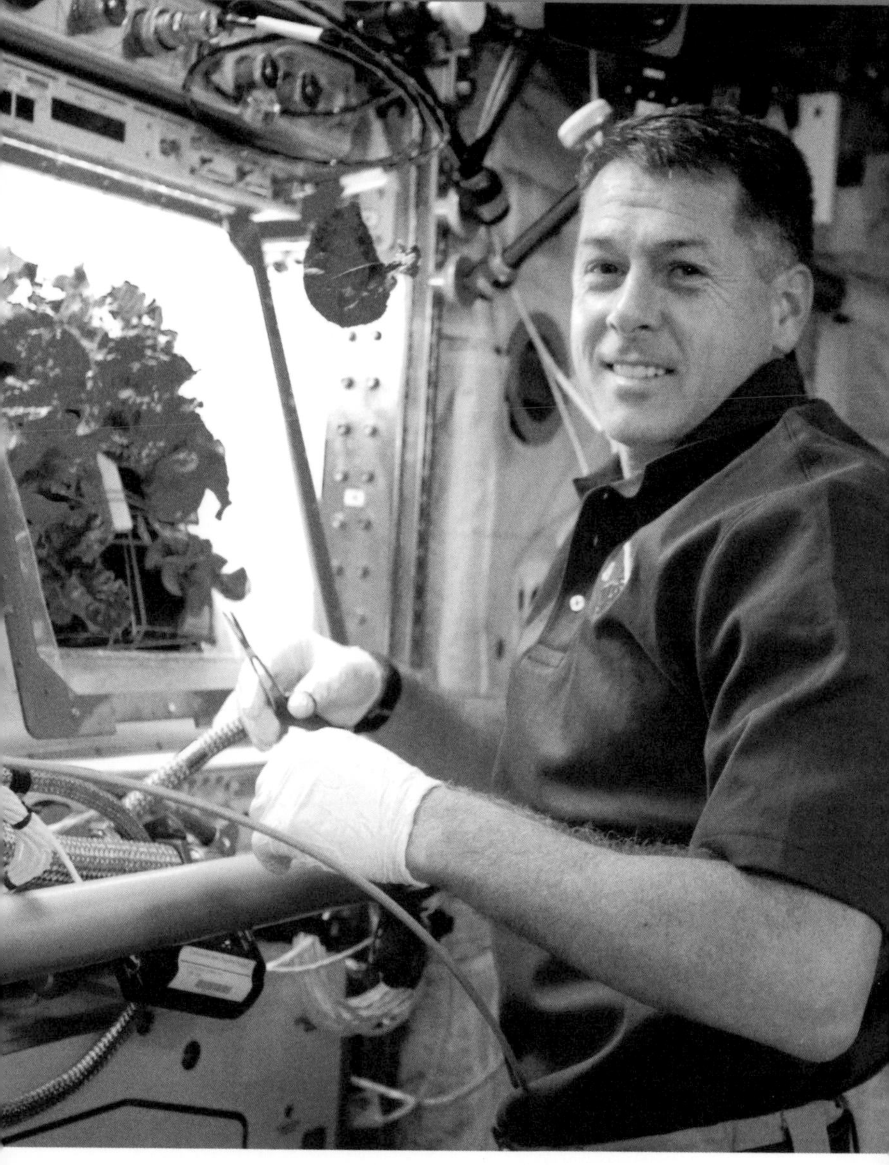

Take a Tour of the 3D Printers

On Mars the 3D printers are as vital as the oxygen and food. Space is at a premium in the Martian habitations and it makes no sense to stockpile everything you could ever need. Instead a computer library holds the designs for clothes, tools, and materials and a 3D printer fabricates them to order. When no longer needed, the materials are recycled back into the system for reuse.

Should some emergency take the colony severely off script, and the necessary tool or part turn out not to be available in the library, one can be custom designed on Earth by a team of experienced engineers and beamed to Mars for download. This easily beats the weeks' to months' travel time of posting a part off by conventional rocket. It can be there within the hour. The 3D printers are opened on the first Sunday of every month for public tours. Every visitor gets a souvenir trinket from one of the machines.

There is also an exhibition delivered via virtual reality headset covering the history of 3D printing in space, starting in 2014 when NASA emailed a design for a wrench to the International Space Station where it was printed by astronaut Barry Wilmore (who is pictured right holding the tool).

Things to Do

The more we do, the more we can do; the more busy we are, the more leisure we have.

Dag Hammarskjöld

Martian Astronomy

Skywatching from Mars is a real treat. While the same constellations are visible, there are many unusual sights without parallel on Earth. You can clearly see Earth in the morning or evening sky as a tiny blue dot. At first you might mistake it for a star, but everyone and everything you've known is bundled up in that speck. The moon is visible alongside it, although it will disappear occasionally as it

A distant view of Earth and its moon from the *HiRISE* camera on NASA's *Mars Reconnaissance Orbiter*

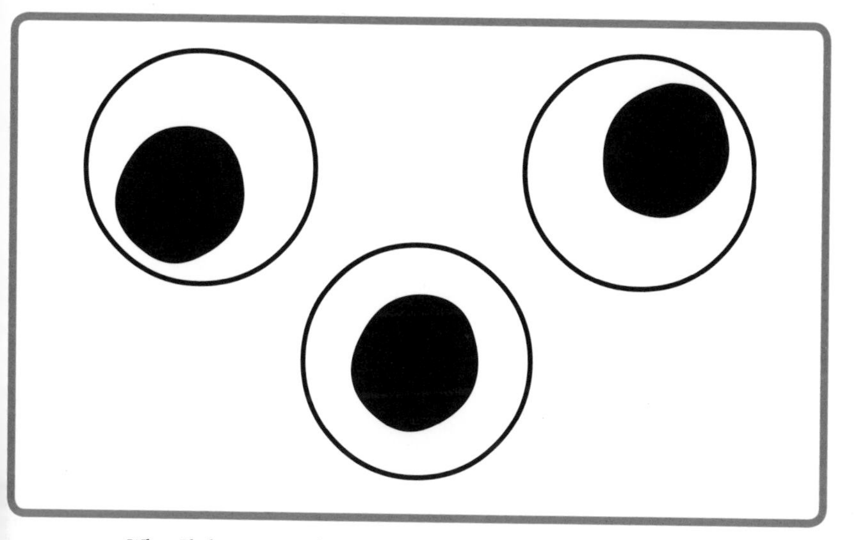

When Phobos passes in front of the sun, it is not big enough to cause a total solar eclipse (as when our moon passes in front of the sun): in images of an eclipse sent back from Mars, Phobos appears about half the size of the sun

moves in front of or behind Earth. Telescopes reveal our planet as a crescent with some vague surface detail. Seeing home up close like this is a real morale booster for homesick travellers.

One of the moons of Mars – Phobos – makes quite a sight, too. It orbits the Red Planet in less than eight hours and so it rises and sets twice a day. It has phases like The Moon, but they change every hour due to its swift passage across the sky. Phobos regularly passes in front of the sun, creating partial solar eclipses in the process. Stickney crater (which was named after Asaph Hall's wife, as we have seen) is clearly visible to the unaided eye. Depending on Mars's position around the sun, Martian astronomers can also get a closer look at Jupiter and Saturn than their terrestrial counterparts.

Visit the *Beagle* 2 Museum

Christmas Day 2003 was supposed to be a momentous day in the history of British space exploration. The European Space Agency's *Mars Express* mission had carried the British *Beagle* 2 lander to the Red Planet. It was named after HMS *Beagle* – the ship that took Charles Darwin to the Galapagos Islands to learn more about life on Earth.

On board its namesake were astrobiology experiments designed to search Mars for signs of life. Mission scientists watched on nervously as the lander headed for the Isidis Planitia. However disappointment was soon to follow. *Beagle* 2 was never heard from again. The mission was declared lost in February 2004 and it drifted out of people's minds.

However, in January 2015, images from NASA's *Mars Reconnaissance Orbiter* revealed it had made it to the surface intact. Two of its four solar panels had failed to deploy, and this had made it unable to communicate.

Today the landing site has been preserved as a museum, with the untouched *Beagle* 2 taking centre stage in the grand atrium. It serves as a reminder that landing on Mars is difficult: its galleries chart the history of our attempts to explore the Red Planet from the earliest probes to the first human missions.

Artist's impression of the *Beagle 2* lander on the surface of Mars

Dune Buggying

This is one for the adrenaline junkies among you. One enterprising company has constructed a dune buggy by first repairing and then re-purposing a faulty rocket engine. They offer thrilling rides through The Namib Dune, part of the Bagnold Dune Field on the

Holiday snaps: a selfie taken by *Curiosity* at Namib Dune

The *Apollo 17* lunar buggy

north-western flank of Mount Sharp inside Gale Crater. The thick outer shell of the buggy is reinforced, offering protection from micrometeorites and radiation. The area was first explored by NASA's *Curiosity* rover in 2015.

On Earth, sand dunes fall into two main groups – small ones with ridge spacings of around ten centimetres (4 inches) and large ones hundreds of metres apart. The unique dunes near Mount Sharp sit somewhere in between with three metres (10 feet) between each. They have been sculpted by Martian winds over many years. The dune field is named after British military engineer Ralph Bagnold (1896-1990) who studied how wind moves sand on Earth.

The much lower gravity on Mars means that the buggy gets a significant amount of "air" as it launches skywards after accelerating up a steep slope. Expect to leave your stomach behind as you fly through the Martian sky. The buggy's large wheels and wide tyres ensure a safe landing.

Aurora Field Trip

Aurorae on Mars are one of the planet's most ethereal and elusive features. On Earth, our magnetic field funnels electrical energy from space towards the poles, thereby generating northern and southern lights in narrow regions known as auroral ovals. However, Mars no longer has a global magnetic field.

A magnetic field remains only in isolated regions of the planet, particularly in the highlands of the southern hemisphere. Several companies run expeditions to these areas in the hope of spotting auroral activity. Don't expect sweeping, vivid green curtains of light as on Earth. Instead, Mars's aurorae appear only in ultraviolet light – beyond the limits of human eyesight. Tour operators will offer you a pair of ultraviolet goggles to allow you to see them. The displays only last for seconds and past auroral activity at a given site doesn't guarantee a repeat show.

Tour operators will check the latest solar weather forecast to estimate the likelihood of increased auroral activity. Be prepared for the possibility that no trips will run during your time on Mars. However, should your expedition go ahead, you must stay in the protective pods provided, as increased auroral activity is linked to solar flares and coronal mass ejections that pose a significantly heightened radiation threat.

The magnetic field lines of Mars

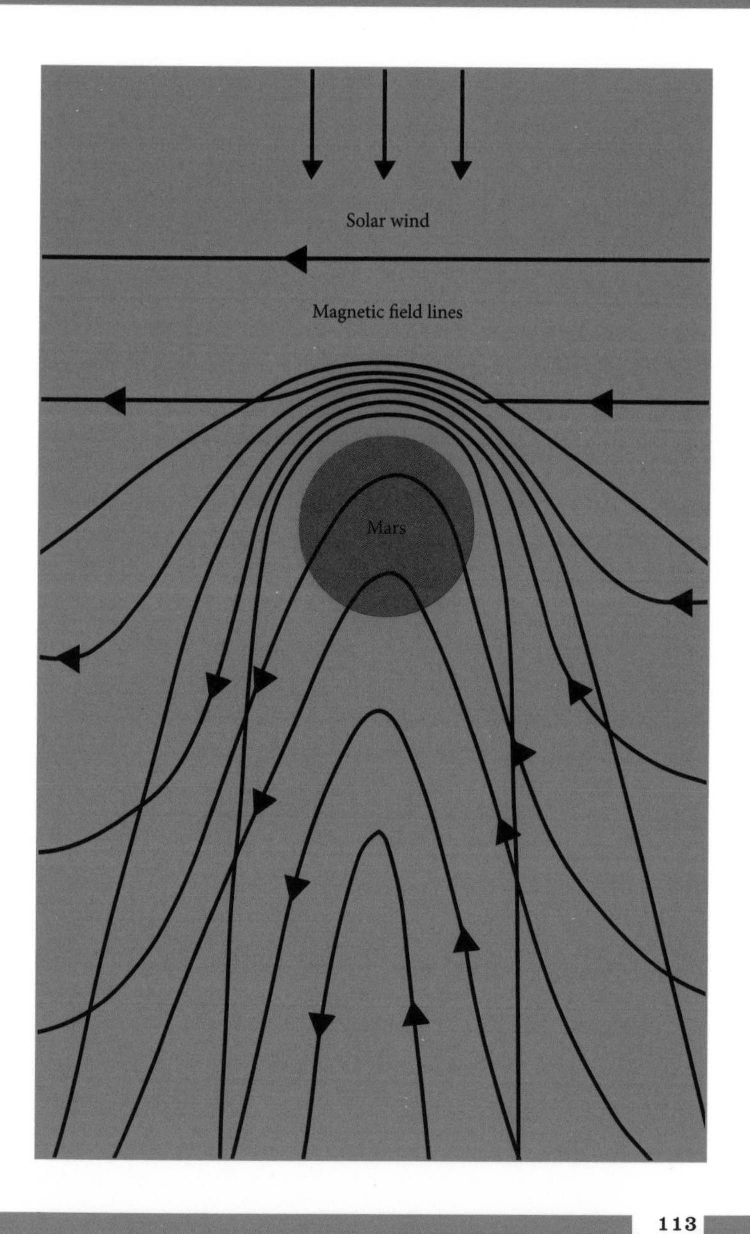
Solar wind
Magnetic field lines
Mars

Sunset Picnic

Fancy a little romance during your vacation? Many of the luxury hotels offer a pre-prepared romantic picnic for two to be enjoyed during a spectacular Martian sunset, all from the comfort of a clear protective dome to shield you from the elements.

The first thing to remember about sunsets on Mars is that you are further from the sun than you are on Earth. So our star appears about 65% of the size it does from home and is only about 40% as bright as it is when seen from the Earth.

Nor should you expect the sunsets to be distinctly red, as they are on Earth. On Mars they tend to take on an eerie blue hue as the ever-present dust in the Martian atmosphere scatters the light differently to an Earth sunset.

Nor are Martian sunsets quick. Once the sun has set on Earth, darkness envelops us pretty rapidly. By contrast, on Mars that same dust high up in the atmosphere helps to bounce the sun's light towards you for up to two hours after it has disappeared over the horizon. This means you get all the more time to enjoy your picnic with the one you love!

Sunset in the Gusev Crater, photographed by NASA's Mars Exploration Rover *Spirit*

Skiing

The slopes of the Martian ice caps offer some of the most spectacular skiing vistas in the solar system. However, the sport here is fundamentally different to the experience on Earth.

For starters, you have to ski in a spacesuit with an oxygen supply. Pit-stops along the route are necessary to restock on oxygen, keeping your weight to a minimum and maximizing your aerodynamic shape. Martian gravity is three times weaker than Earth's, so you need to slalom down slopes that are three times higher in order to attain something close to terrestrial speeds.

On Earth, the pressure exerted by your weight helps to melt the snow under your skis to provide a thin layer of water to glide on. Martian skis are narrower and shorter, which maximizes the amount of downwards force on the ice in order to replicate this effect. Terrestrial skis are normally black so as to absorb solar energy and assist with melting. Martian sunlight is simply too weak to achieve this.

On some of the trickier slopes there is no water ice, only dry ice. Solid carbon dioxide will turn straight into a gas and so won't provide a liquid layer to ski on. Larger skis with the underside coated in a special wax provide the necessary lubrication.

Snowy hills in late spring. This image shows south-facing slopes within a crater.

Play Pitch and Putt

Humans have history when it comes to extra-terrestrial golf. When Alan Shepherd went to the moon with *Apollo 14*, he smuggled the head of a six-iron and two golf balls with him. He attached the head to the handle of a lunar sample scoop to make a makeshift club and hit the balls on the lunar surface.

Alan Shepard carries out experiments before his famous lunar golf shots

Mikhail Tyurin

Then, in 2006, cosmonaut Mikhail Tyurin hit a golf ball during a spacewalk from the International Space Station (ISS) in a publicity stunt for a Canadian golf company. Like Shepard, he used a six-iron, while ISS commander Michael Lopez-Alegria had to hold his legs during the shot in order to give him stability.

A full round of golf on Mars is out of the question for several reasons. The weaker gravity means you could drive the ball much further than on Earth and the meagre air resistance provided by the thin Martian atmosphere would boost your shot length even more. A professional golfer could drive over 800 yards (730 metres) on Mars, so the courses would be enormous and a waste of precious space. However, an indoor pitch and putt course has been built to allow you to experience the fun of golf in a different gravitational setting.

Sandboarding

Think surfing requires water? Think again. On Mars you can surf down the sand. The activity was inspired by a naturally occurring phenomenon called linear gullies.

These tracks in the sand can be up to a mile (1.6 km) long and ten metres (33 feet) wide. They are formed when a block of dry ice breaks off in Martian spring and surfs down a slope on a layer of sublimating carbon dioxide.

The sand is much finer than you'd find in Earth's sand dunes. You'll be issued with a sandboard, very similar to the snowboards you get in ski resorts on Earth where your feet are strapped in.

Part of the Bagnold Dunes, along the northwestern flank of Mount Sharp, a popular sandboarding destination

The bottom of the board is covered in a paraffin-based wax to emulate the sublimating CO_2 and help you glide over the sand.

Wider boards similar to toboggans are available for less confident adventurers who'd prefer to be seated. Sandboarding offers a great summer alternative to skiing when the Martian icecaps are considerably smaller than in the winter.

On Earth the same activity is offered in the dunes of Bolivia, Dubai, and Australia. However, as with skiing, you'll have to start your run from a much greater height to achieve the same speed as on Earth because Mars's gravity is considerably weaker.

Rock Climbing

If the lower Martian gravity makes skiing more difficult, it certainly makes rock climbing easier. However, the life support and gear you'll need to carry with you will nullify some of this advantage. The fact that a lot of Martian rock is comprised of volcanic basalt is especially good news for climbers.

As the lava cooled, it cracked to form joints that are often arranged in columns. They provide excellent steps for parts

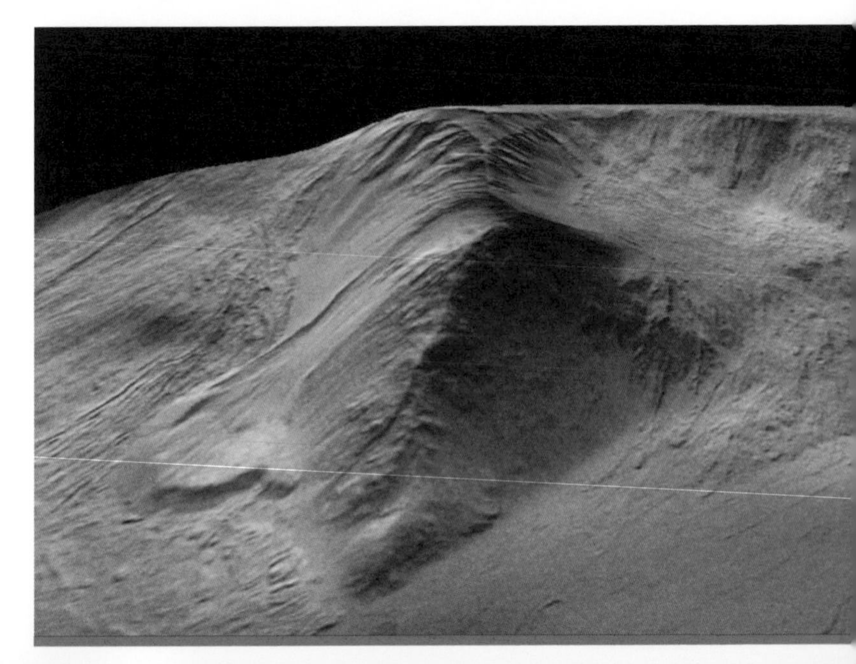

of any ascent. Gas bubbles also formed in the cooling lava, producing features called vesicles that are the perfect finger holes to grip on to as you climb. While you can wear your normal spacesuit for mountaineering on Mars, be sure to take plenty of repair patches in case you snag your suit on a jagged rock.

You will, however, need a complete change of gloves and boots. Climbing requires the hands and fingers to be highly dextrous and your normal gloves are too restrictive. In a similar manner, your standard issue boots don't generally have sufficient grip to contend with serious climbing.

The best climbing on the planet is to be found in the Valles Marineris and in the 60 kilometer-wide (37 mile-wide) caldera at the top of the extinct Olympus Mons volcano (pictured below).

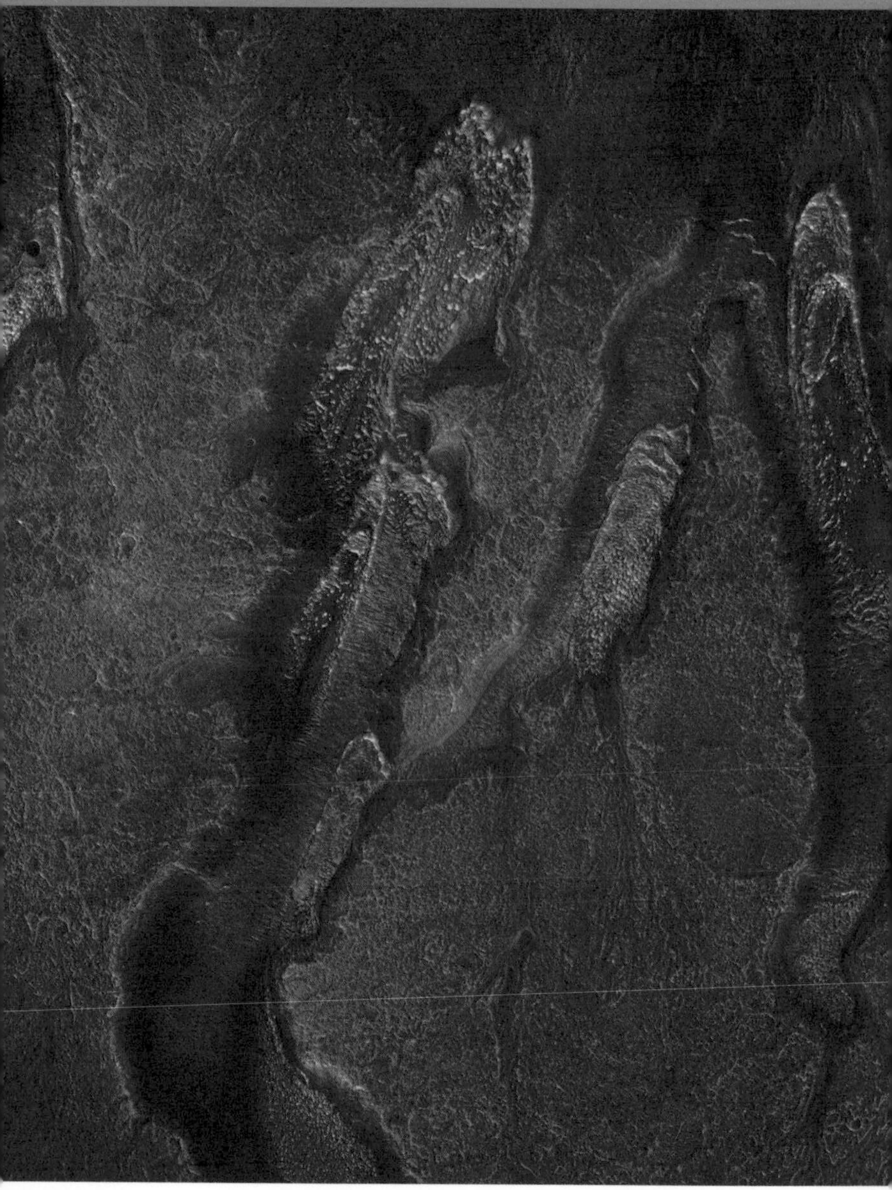

Glacier Trekking

You don't have to venture to the frigid Martian poles for winter activities. Mars has glaciers – slow moving rivers of ice – in bands between 30 and 50 degrees above and below the equator. They were initially spotted using radar measurements from orbit, as they were otherwise hidden under vast dust deposits.

Some of the dust has been cleared for adventurous tourists to experience walking on parts of a Martian glacier. However, this has to be highly managed as the dust on top is what stops the ice from disappearing. The regions open to visitors are constantly rotated to preserve the glaciers for the long term. Water is a precious resource on Mars, after all. A total of 150 billion cubic metres of water is locked up inside Mars's glaciers. Were they all to melt, the resulting flood could cover the entire planet to a depth of one metre.

More experienced adventurers should attempt to tackle the unique landscape of the glaciers that are found in the Ismenius Lacus quadrangle. The rugged terrain of this area consists of furrows, ridges, and folds that pose a real challenge to hikers. The area around Moreux crater is especially breathtaking and has been extensively altered by past glacier activity.

These glaciers, in the south-facing wall of a crater, are unusual as they have bright highlights, probably made up of shiny dust and dark sand

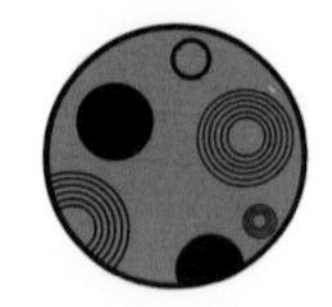

Underground Spa

The surface water of Mars may long since have vanished, but that doesn't mean it is a completely dry planet. Residual heat from the planet's core is enough to sustain giant underground aquifers which are filled with ancient rains that once poured down from the Martian sky.

Heated by contact with high temperature rocks, hot springs bubble up in some parts of the volcanic lava tubes of Pavonis Mons close to the hotel caverns. Like thermal spas on Earth, the water contains dozens of restorative minerals.

Day spa packages are available during which you can indulge in pampering and relaxation. Entry includes one of two additional treatments. The most popular option is the mud treatment, similar to the Moroccan Rhassoul clay treatments available on Earth. The clay comes from in and around large Martian impact craters formed billions of years ago. Applying a Martian mud mask draws out oils from the skin and unclogs pores.

Hot stone massages are also available using smooth and rounded basalt rocks excavated from the slopes of the nearby Tharsis Montes trio of volcanoes. The high iron content of Martian rocks means they can hold heat for a long time, making them the ideal choice for getting rid of those aches and pains.

Mars in Popular Culture

No one would have believed in the last years of the nineteenth century that this world was being watched keenly and closely by intelligences greater than man's and yet as mortal as his own.

H . G. Wells

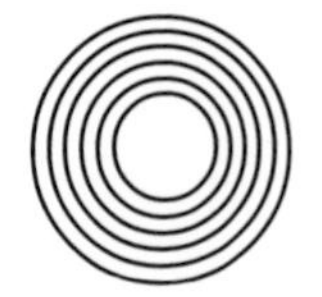

The War of The Worlds

Think of fictional portrayals of the Red Planet and the chances are that H.G. Wells's 1898 classic will come to mind first. The plot sees southern England invaded by extraterrestrials from Mars. The Martians – the size of bears and with tentacles around their mouths – initially struggle to breathe Earth's atmosphere. So they enclose themselves inside huge, armoured tripods instead, and set off on a rampage through Surrey, incinerating people with their heat-rays and poisonous black smoke.

Humanity fights back, but London eventually falls and it looks as if the Martians have succeeded in conquering Earth. Martian red weed is seen growing everywhere. However, the invaders succumb to an infection that their alien biology cannot withstand.

This powerful story has been adapted and retold many times. In 1938, Orson Welles famously narrated and directed a radio broadcast that was so vivid many listeners tuning in part way through were convinced the Martians were really here. In the 1970s, Jeff Wayne turned the story into a musical album with Richard Burton providing the voice of the narrator. Tom Cruise starred in the 2005 Hollywood adaption, playing the lead character, who was named Ray Ferrier. (In Wells's original text the main character remained nameless.)

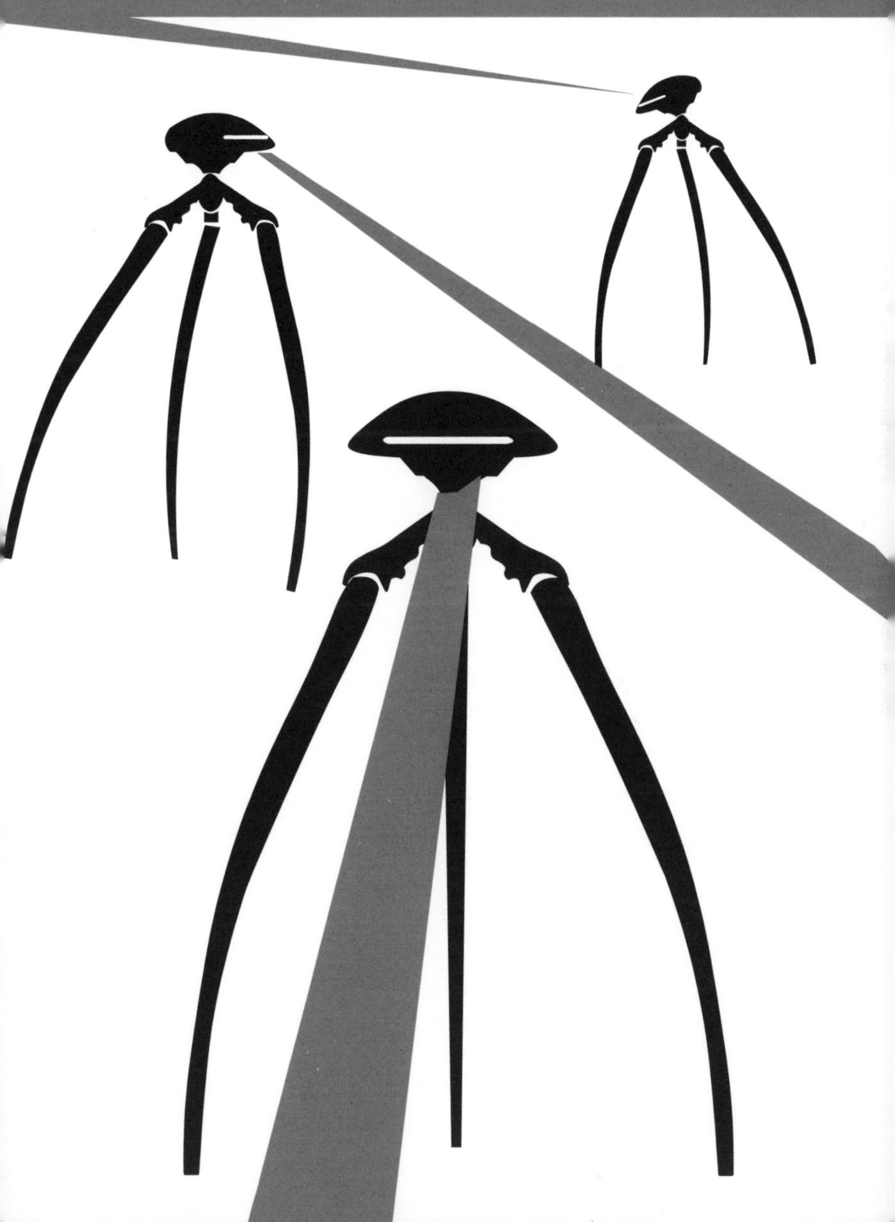

Mars Attacks!

A unique blend of comedy, kitsch, and sci-fi, Tim Burton's 1996 film *Mars Attacks!* is a parody of science fiction B-movies based on a cult trading card game of the same name. The plot sees hundreds of Martian ships descending on the Earth and invading the planet. Several prominent characters are captured by the

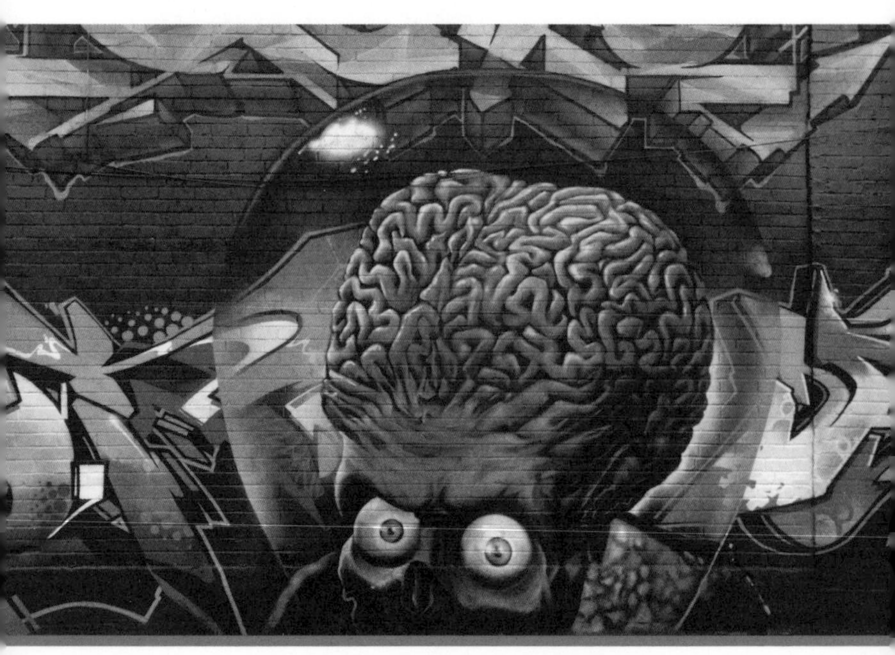

Street Art in Montreal based on *Mars Attacks!*

Martians and experimented on. In one memorable scene, the newscaster played by Sarah Jessica Parker has her head swapped with that of her pet chihuahua. In another scene, Pierce Brosnan's character Professor Donald Kessler awakes to find his head has also been severed and he is surrounded by his own dismembered body parts. Later the two disembodied heads profess their love for each other. The singer Tom Jones even makes a cameo appearance when his gig in Las Vegas is suddenly interrupted by the Martians firing their laser guns into the crowd.

The Martians appear to be winning the war against the Earthlings, invading the White House and even killing the President. Then, as in H.G. Wells's *The War of the Worlds*, their Achilles' heel is discovered – this time it is the song *Indian Love Call* by Slim Whitman, which makes their heads explode.

The original series of trading cards was published in 1962. Created by science-fiction artists Wally Wood and Norman Saunders, each card depicts a scene in the invasion of Earth by vicious Martians trying to colonize the Earth in order to escape from their own doomed planet.

The series was reissued in the 1980s with an expanded merchandise range, and Tim Burton (pictured above speaking at Comic-Con) based his movie on it.

The Martian

Directed by Ridley Scott, this 2015 film was based on the novel of the same name by Andy Weir. It won the Golden Globe for Best Picture (Musical or Comedy) and Matt Damon won the Best Actor award in the same category. Set in 2035, the movie opens with a fierce Martian dust storm that suddenly plunges a human crew on the planet into imminent danger.

As they scramble for refuge, Damon's character – Mark Watney, the mission's botanist – takes some flying debris to the chest and is left for dead as everyone else jumps aboard an escape rocket and blasts off back to Earth.

Watney survives, makes it safely back to the habitation pod and eventually realizes he has to wait it out on Mars alone for four years before anyone can rescue him. NASA engineers discover he's alive

Hydroponics is a way of growing plants without soil, by exposing the roots directly to a mineral solution. This method would be used both on the flight to Mars and in the surface colonies.

when images from orbit clearly show human activity. The crew en route to Earth turn around and go back for him.

The movie is notable for the wit, humour, and resilience of Watney and his cheerful resignation to his solitude. He keeps a video diary, grows potatoes for food using his own excrement as fertilizer and commandeers a camera salvaged from the old *Pathfinder* mission to communicate with Earth.

This ingenious spot of hydroponic improvization is actually technically feasible although there are one or two complications: Martian soil contains perchlorates, a type of salt that would be toxic to consume. However, given a supply of water, which we know is available on Mars, the perchlorates could simply be washed out.

The other danger of using excrement to add biological material to Mars soil is that you would be consuming potentially hazardous pathogens. As long as Watney is using his own excrement, he is only exposed to his own pathogens, so is not endangered. The book goes into more detail than the film, explaining that when Watney expands his system by using the frozen waste left behind by the other members of the crew, the fact that this has been frozen and dessicated means that any pathogens have been destroyed.

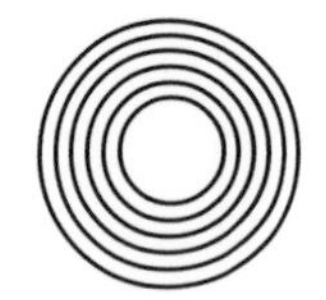

The Mars Trilogy

Kim Stanley Robinson's acclaimed trilogy of books imagines our long term attempts to transform Mars into a planet fit for long term human habitation.

The first instalment – *Red Mars* – begins in 2026 with our initial colonization by a crew of 100 people. An outpost is also established on Phobos. Holes are drilled into the Martian surface in order to release heat and nuclear devices are detonated underground to liberate water. Back on Earth, humans end up embroiled in another world war.

Green Mars – the second book – picks up the action at the beginning of the 2100s. Disorder has broken out as different countries and factions vie for control of Mars. Orbital mirrors have been installed to warm the planet and the rebels consider sabotaging them. Mars's northern ice cap has been melted to provide more water. On Earth, the constantly rising sea levels have caused catastrophic floods.

By the final book, *Blue Mars*, the planet has enough of an

atmosphere for liquid water to be stable on the surface and to form seas and rivers. There is much illegal immigration from the stricken Earth to this newly terraformed Mars. This forces humanity into terraforming other parts of the solar system too, including Venus and even some of Jupiter's moons.

It has long been rumoured that the series will be adapted for television – at one point James Cameron held the rights and there has been a significant amount of development work carried out over the years. If this comes to fruition, the strong emphasis on the characters involved in carrying out the process of terraforming the planet will make it fascinating viewing.

Total Recall

It's 2084 and Arnold Schwarzenegger's character Douglas Quaid is having recurring disturbing dreams about Mars. In the 1990 film, the Red Planet is a mining outpost facing a rebellion due to the governor's monopoly on breathable air.

Stuck on Earth, Quaid visits a company that can implant false memories into his brain in order for him to experience having an adventure on Mars without making the arduous journey there.

This triggers a chain of events that sees Quaid revealed as an actual secret agent and sent to the Red Planet for real. He checks into the Hilton Hotel on Mars under a pseudonym and ends up in Venusville – the planet's red light district – which is full of people who have been mutated by exposure to radiation. The rebels are searching for a rumoured alien artefact hidden somewhere in the mines called a turbinium reactor. It can supply all the breathable air required and shatter the contentious monopoly.

Both Jaws actor Richard Dreyfuss and Patrick Swayze were considered for the role of Quaid before Schwarzenegger. The film was remade in 2012 starring Colin Farrell.

Total Recall was based on the short story "We Can Remember It For You Wholesale" by sci-fi writer Philip K. Dick. As with a number of other film adaptations of Dick's work, the screenplay takes only a small part of the plot from the source. The story begins in exactly the same way, with Quaid wistfully wanting to go to Mars and having false memories inserted, only to be informed that there has been a problem with the

operation, as he has real but suppressed memories. However, the plot then develops in a more philosophical vein. Quaid, who is more of a Walter Mitty character, in contrast to Schwarzenegger's turbocharged macho man, ends up with two contradictory memories and Dick explores the problem of how we know what our true memories are.

In the end, Quaid's superiors offer him a deal – in order to erase his memories of Mars, as they need to do, they have to implant an alternative memory in which Quaid's dreams are fulfilled. This is so that he doesn't just start yearning for Mars again and end up going back through the same loop.

The memories that are inserted will vindicate Quaid's wishes to be important by convincing him that he is the most important person in the world, as he is the only human on earth who can ward off an alien invasion.

However, once his superiors insert this false memory, he once again discovers that the false memory is the truth, and that he really is protecting the world from such an invasion.

Marvin the Martian

First appearing in the 1948 Looney Tunes cartoon *Haredevil Hare*, Marvin the Martian is an adversary of Bugs Bunny. He is quiet, yet sneaky and destructive – a deliberate contrast to Bugs's other foe Yosemite Sam. Wearing a Roman uniform in a nod to the god Mars, the traditional centurion's feather plume is replaced with the head of a broom. He has a black head with two eyes but no mouth, yet is somehow able to speak.

Bugs Bunny regularly thwarts Marvin's attempts to blow up the Earth with a pew-36 explosive space modulator. Frustrated, Marvin is often heard to ask "where is the kaboom?" His justification is that Earth blocks his view of Venus. Throughout his appearances in various cartoons he has a love interest – Queen Tyr'ahnee. However, she does not love him back. He is also often followed by his dog K-9, a green version of Mickey Mouse's pooch Pluto. In the Looney Tunes film *Space Jam* with Michael Jordan he is the basketball referee – the only impartial character between the Looney Tunes and the alien opposition. NASA scientists are clearly fans: the official launch patch for NASA's *Spirit* rover had Marvin the Martian as its main image.

People of Note

By three methods we may learn wisdom: First, by reflection, which is noblest; Second, by imitation, which is easiest; and third by experience, which is the bitterest.

Confucius

Giovanni Schiaparelli

The astronomical career of Giovanni Schiaparelli (1835–1910) took him from his Italian homeland to Germany and Russia before he returned home to start a forty year stint at the Brera Observatory in Milan. He was one of the first to spot an apparent series of lines across the Martian surface that would gain fame as artificial canals.

His astronomical work also saw him discover the asteroid 69 Hesperia in 1861 and link the annual Leonid meteor shower to the comet Tempel–Tuttle. He also made some detailed observations and drawings of Mercury and Venus and attempted to calculate their rotation periods.

His work on comets was influenced by his tutor in Berlin, the famous comet discoverer Johann Encke. In recognition of his

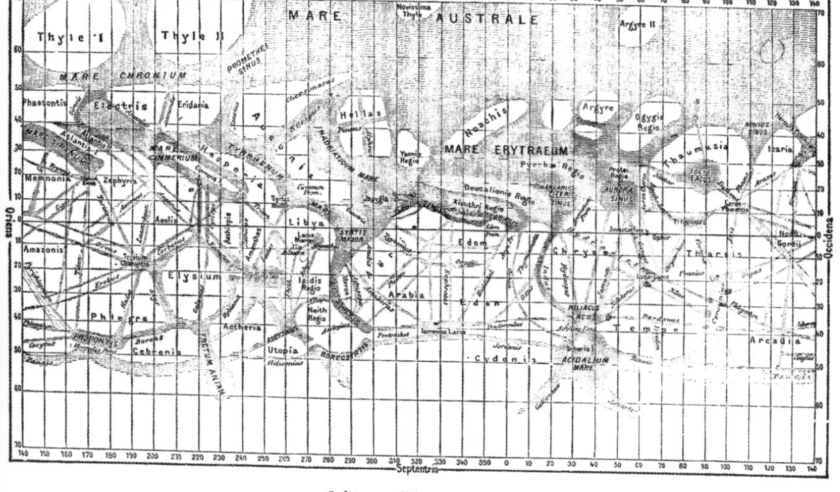

Schiaparelli's 1889 map

significant contributions to astronomy there are craters on the moon and Mars named after him, along with a region of Mercury – a planet he studied extensively. The asteroid 4062 Schiaparelli is also named after him.

The European Space Agency and Russian Space Agency, Roscosmos, attempted to land the *Schiaparelli* probe on Mars in October 2016. It was a test-run for the forthcoming ExoMars mission. However, the landing sequence failed to execute properly and the lander smashed into the Martian surface at over 500 kilometers an hour. It was a reminder that landing on Mars still presents a huge technological challenge.

Percival Lowell

When it comes to Mars, Percival Lowell's name is perhaps the most notorious. Born in Boston in 1855, he founded the Lowell Observatory in Flagstaff, Arizona – a site he'd deliberately chosen for its dark skies. One of the finest observatories of its day, Pluto was discovered there in 1930 some 14 years after his death.

However, it is with the infamous canals on Mars that Lowell is inextricably linked. He studied the Red Planet from Flagstaff for a fifteen-year period starting in 1894 and published several books with highly detailed maps of the supposed Martian canal system. In their pages he explicitly argues in favour of intelligent life forms on Mars.

Today we not only know that the canals aren't there, but also that there isn't even a network of dark lines on the surface of Mars. So what was Lowell drawing? Even through a big telescope Mars appears small, its surface features hazy. Perhaps Lowell was sketching what he wanted to see. Given that he also noted spokes on the surface of Venus, some modern astronomers have suggested that the way his telescope was set up meant he was really seeing reflections of the blood vessels in his own eyes.

Walter Maunder

A British astronomer working at the Royal Observatory Greenwich, Walter Maunder (1851–1928) was an early sceptic of the Martian canals. He famously said "We have no right to assume, and yet we do habitually assume, that our telescopes reveal to us the ultimate structure of the planet." He showed that irregular dark patches on Mars could appear to have straight edges if straining to see them through the telescopes of the day.

Maunder is more famous for his work on the sun. His main job at Greenwich was to observe sunspots on the surface of our star. In studying historical records of sunspot numbers he noticed a fallow period between 1645 and 1715 that astronomers now call the Maunder Minimum. This partly coincided with a period known as the Little Ice Age – a time when European temperatures dropped below their long term average.

Maunder married Annie Russell, one of the "human computers" employed at Greenwich to assist with calculations. Victorian social convention meant she was obliged to give up her job on marriage. However, she continued to make important contributions in her own right, although her gender often prevented her from having her name attached to her work.

Asaph Hall

Asaph Hall (1829–1907) was an American astronomer famous for discovering Mars's two moons Phobos and Deimos during the planet's close approach in 1877.

The son of a clockmaker, money was tight as a child and Hall initially trained as a carpenter. Largely self-taught, he took maths lessons from the abolitionist and suffragette Angeline Stickney, two years his senior, and the pair later married.

Astronomy turned out to be his true passion. He was a gifted celestial mathematician and in 1862 he was appointed to the post of Assistant Astronomer at the US Navy Observatory (USNO) in Washington D.C. In 1875 he was put in charge of the USNO's 26" telescope – the largest refracting telescope in the world at the time, and the one he used to find the moons of Mars.

He also studied Saturn and its own natural satellites, most notably the oddly behaved Hyperion, calculated the distances to stars, and tracked the motions of stars inside clusters. Hall retired from USNO in 1891 and was soon inducted into the prestigious French *Légion d'Honneur* before taking up a teaching post at Harvard. Craters on the moon and Mars are named after him, as is the asteroid 3299 Hall.

Carl Sagan

Carl Sagan (1934–1996) was a hugely influential and popular American astronomer. Born in Brooklyn, New York, he was perhaps best known for bringing astronomy and space science to the general public. His TV series *Cosmos* endeared him to millions around the world with his poetic words and instantly recognizable voice. He also wrote the sci-fi novel *Contact* on which the 1997 film starring Jodie Foster was based.

He made important contributions to science, too, playing a key role in many of the missions to explore Mars in the second half of the twentieth century. Part of the imaging team behind the *Mariner* 9 probe, he also helped select the landing sites for the *Viking* 1 and 2 landers. Yet even in the 1970s he was thinking big. He made no secret of his desire to see rovers sent to Mars to drive about and explore more of the Red Planet.

He died of cancer-linked pneumonia on December 20 1996, less than three weeks after NASA launched the *Pathfinder* mission to Mars, but several months before *Sojourner* drove out for the first time in July 1997. Its landing site was renamed the Carl Sagan Memorial Station in his honour.

Adam Steltzner

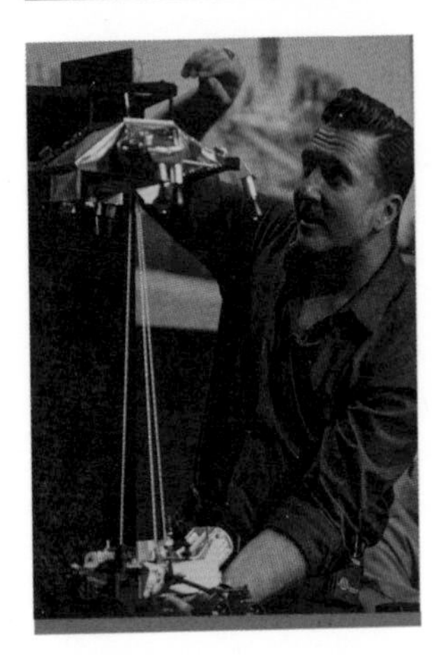

Born in 1963 in Alameda County, California, Adam Steltzner works for NASA's Jet Propulsion Laboratory. Famous for his rock star look – complete with snakeskin shoes and Elvis haircut – he has worked on the *Pathfinder*, *Spirit* and *Opportunity* Mars missions. He was also the lead engineer for the hair-raising sky crane descent phase of the *Curiosity* rover that became known as "The Seven Minutes of Terror."

His career has been the subject of much media interest because his path to NASA was far from typical. Steltzner's father was the heir to a massive American food company, but, by his own admission, Adam was more interested in sex, drugs, and rock 'n' roll and performed in several bands.

Inspired by looking at the night sky one night, he took an astronomy class and it changed his life. By 1991 he had a degree from CalTech and he completed his PhD in engineering in 1999. Along with his work on the Mars missions, he was also involved in the *Galileo* mission to Jupiter and the *Cassini* mission to Saturn. He regularly appears on TV and radio and was inducted into the National Academy of Engineering in 2016.

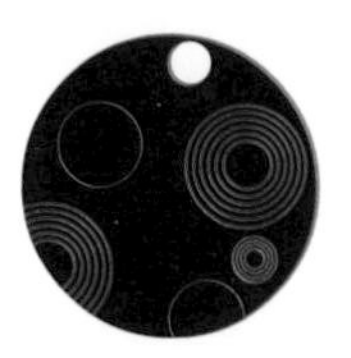

Robert Zubrin

Born in Lakewood, Colorado in 1952, Robert Zubrin is a long time advocate of human missions to Mars. In 1998 he founded the international, not-for-profit Mars Society with the aim of drumming up interest in privately funded trips to the Red Planet.

They hold an annual International Mars Society Convention and have set up Mars training stations in Hanksville, Utah and Devon Island (an uninhabited region of the Canadian Arctic).

With a PhD in nuclear engineering, he has written over 200 scientific papers related to the propulsion methods necessary to make such trips a reality and he holds several relevant patents. While working at Lockheed Martin in the 1990s, he devised the *Mars Direct* mission plan – a way to get there at just 12.5% of the cost previously estimated by NASA by creating oxygen, water and rocket fuel from the Martian atmosphere. Zubrin has also talked about the possibility and ethics of terraforming Mars one day in order to turn it into a permanent, habitable home for humanity. Frequently appearing in media discussions about Mars, he is the author of several popular science books including *The Case for Mars* and *How to Live on Mars*.

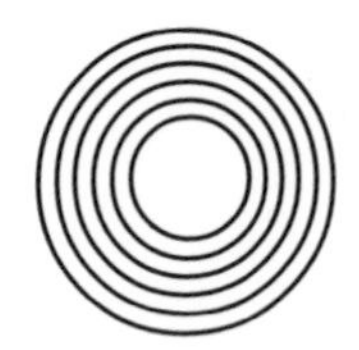

Elon Musk

To many, Elon Musk is the most visionary entrepreneur of the 21st century. Born in South Africa in 1971, he made his fortune after selling an 11.7% stake in online payment service PayPal. He is now CEO of both SpaceX, a private space company, and electric car manufacturer Tesla Inc.

However, his sights are set firmly on Mars. He has stated publicly on numerous occasions that his goal is to build a city on Mars with a million inhabitants in the next fifty years. The cost of a ticket could be just $500,000 – a price tag he believes many reasonably affluent people could afford by selling their homes and other earthly possessions.

It may sound like a pipe dream, but his company SpaceX is already revolutionizing access to space and has won NASA contracts to deliver goods to the International Space Station.

In the coming years they hope to send paying customers around the moon. Space is fast becoming a place not just for the lucky few and highly trained. Soon, everyday people will routinely go into space for leisure. With governments coming under increasing scrutiny from taxpayers, don't be surprised if Musk and SpaceX win the race and beat NASA to the Red Planet.

The Future

The only true happiness is to learn, to advance, and to improve: which could not happen unless we had commenced with error, ignorance, and imperfection. We must pass through the darkness, to reach the light.

Albert Pike

Artificial Intelligence & Virtual Reality

Our exploration of Mars to date has all been robotic because sending people so far away presents many sizeable challenges, not least the food, water, oxygen, and extra protection required. Yet a single human geologist on Mars could probably achieve more in a day than all of our artificial emissaries combined.

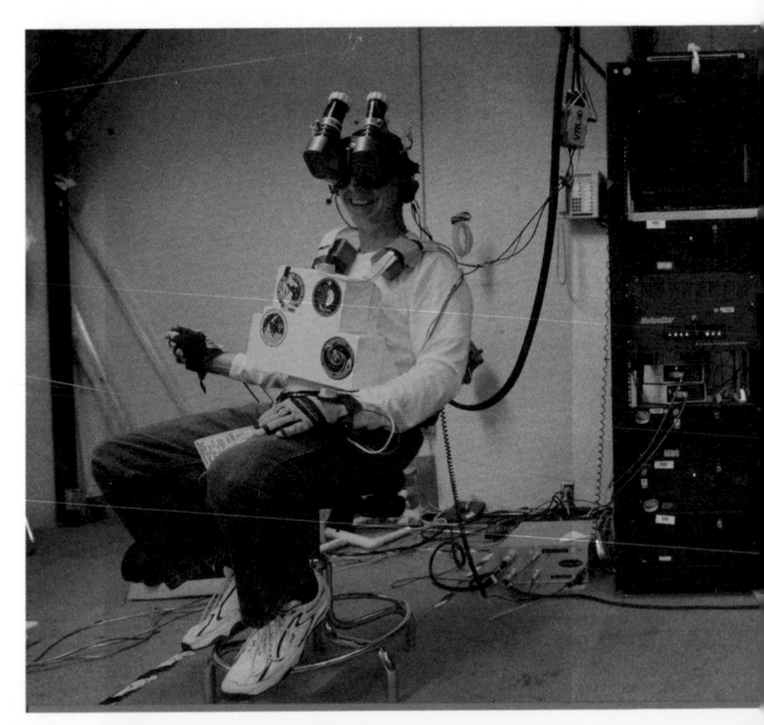

Advances in artificial intelligence and virtual reality might provide a happy halfway house. For instance, androids making their own decisions based on pre-programmed parameters could go a long way towards emulating a field geologist. And they'd need no sleep or sustenance. Any accident would be expensive, but not the tragedy that comes with losing human life. There would be no psychological problems or a longing for home and family to deal with.

Another possibility is that developments in virtual reality might allow a field geologist to control a robotic counterpart from afar. Probably not from Earth because there would be at least a twelve minute delay as signals pinged between the planets.

That kind of latency would prevent the operator from responding in real time. However, controlling a rover from orbit around Mars would eliminate the danger of landing on the Red Planet and the expense and risk of taking off again to return home.

NASA astronauts use virtual reality hardware in the Space Vehicle Mock-up Facility at NASA's Johnson Space Center

Terraforming

Many of the problems with living on Mars stem from the fact that it is a hostile environment very different from the planet our bodies have evolved to cope with. What if it wasn't so different? Thinking long term, we might engineer Mars to be more like home. This process is called *terraforming*.

The good news is that it's a runaway process. Warm the planet slightly and some of the ice will melt, releasing greenhouse

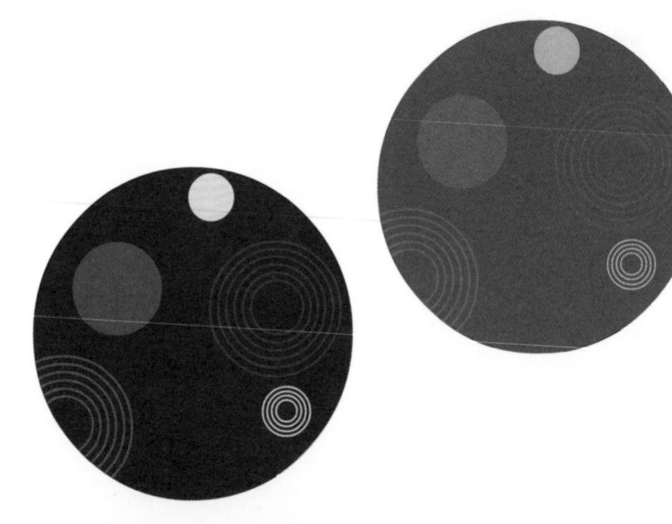

gases that have been trapped for billions of years. They will help raise the temperature further, releasing more gas to provide yet another boost. Eventually water will flow again on Mars, plants can grow outside and humans can breathe without assistance.

The bad news is that it's not a quick process. It might take a thousand years of incremental change to fashion Mars in our image. How might we be able to do it?

Perhaps by installing a large mirror in Mars orbit to reflect extra sunlight down on the ice. A single mirror 240 kilometer across could raise the temperature by 18 degrees. Of course, that level of engineering is far from our current capabilities. Other ideas include crashing an asteroid into the planet or introducing gas-producing bacteria.

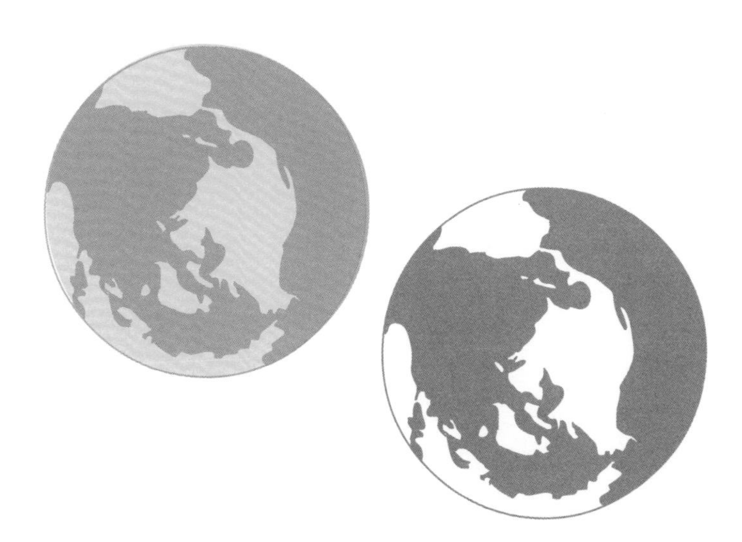

Humans on Mars

Just when will the first person set foot on Mars? It is really a question of how much money we are willing to spend on getting us there. Elon Musk's ambitious plans with SpaceX include reaching the Red Planet with a human crew as early as 2024. NASA's current aim is for people to make the journey sometime in the 2030s. While all eyes are on the US, China is quietly going about its business. Beijing is eyeing up sending their first probe to Mars by 2020 and might be able to steal a march on the competition if they make significant progress over coming years. Perhaps one of their taikonauts (the Chinese word for astronaut) will be the first to reach Mars. The European Space Agency is also drawing up plans, so this century looks a fairly safe bet.

A lot of what happens after our initial trip depends on how easy it proves to overcome some of Mars's unique challenges. Will underground lava tubes be sufficient to shield us from radiation? Will the psychological effects prove too difficult?

It's definitely a question of when, not if. Mars holds such a draw that we will make it there eventually, in the process starting to insure ourselves against threats that could wipe out life on Earth and becoming a two planet species.

Index

Picture credits

All illustrations and images by Diane Law except for the following: Mars (p.7): Nasa/JPL-Caltech / Topography (pages 8-9): NASA / JPL / USGS / Akihiko Hoshide (p.17): NASA / The vomit comet (p.19): NASA / Tim Peake (p.25): ESA/NASA / Space food (p.26): NASA / Hydroponics warehouse (p.27): Bryghtknyght/Creative Commons / Lava tubes (p.32): ESA/DLR/FU Berlin / Astronaut (p.33): NASA / Dust storm (p.34-35): NASA/JPL/MSSS / Cosmic ray environment (p.36): NASA / Meteorite (p.39): NASA/JPL-Caltech/LANL/CNES/IRAP/LPGNantes/CNRS/IAS/ MSSS / Mars, God of War (p.45): Andrea Puggioni/Creative Commons / Telescope (p.49): Public domain, painting by Adriaen van de Venne / HiRISE over Mars (p.50): NASA/JPL / Phobos (p.51): NASA/JPL/University of Arizona / Percival Lowell (p.52) and his drawing (p.53): Public domain / Mars (p.54): NASA/GSFC / Mariner 4 (p.55): NASA/JPL / Olympus Mons (p.56): NASA / Trenches (p.57): Nasa / Carl Sagan (p.58): JPL / Surface

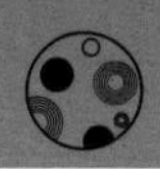

(p59-60): NASA/JPL / Victoria crater (p.61): NASA / Curiosity (p.63): NASA/JPL-Caltech/MSSS / Rover (p67): NASA/JPL-Caltech / Chasma Boreale (p.72): ESA/DLR/FU Berlin / Sand (p.73): HiRISE/MRO/LPL(U. Arizona): NASA Tharsis Montes (p 75): NASA / JPL-Caltech / Arizona State University / Gale crater (pp.76-77): NASA/JPL-Caltech/MSSS / Terra Sirenum (p.78): NASA/JPL-Caltech/University of Arizona / Srytis Major (p.83): NASA / Hesperia Planum (p.85) and Meridiani Planum (p.86): ESA/DLR/FU Berlin / Garni crater (pp.88-89): NASA / JPL-Caltech / Arizona State University / Orcus Patera (pp.90-91): ESA/DLR/FU Berlin (G. Neukum) / Olympus Mons (p.93): NASA/MOLA Science Team/ O. de Goursac, Adrian Lark / Valles Marineris (pp.94-95) NASA/JPL / Hellas Basin (p.96): MOLA Science Team / Phobos (p.98): NASA / Greenhouse (p.103): NASA / Barry Willmore (p.104): NASA / Earth and moon (p.106): NASA/JPL/University of Arizona / *Beagle* 2 (p.108): ESA/Denman productions ESA/Denman Productions / *Curiosity* (p.108): NASA/JPL-Caltech/MSSS / *Apollo* 17 (p.109): NASA / Sunset (p.113): NASA/JPL/Texas A&M/Cornell / Snowy slopes (p.116): NASA/JPL/University of Arizona / Alan Shepard (p.118): NASA/Edgar Mitchell / Mikhail Tyurin (p.119): NASA / Bagnold Dunes (p.120): NASA/JPL-Caltech/MSSS / Olympus Mons (p.122-123): ESA/DLR/FU Berlin, CC BY-SA IGO 3.0 / Glaciers (p.124): NASA/JPL/ University of Arizona / *Mars Attacks!* (p.130): Shutterstock / Tim Burton (p.131): Gage Skidmore / Kim Stanley Robinson (p.132): Gage Skidmore / Schiaparelli (p.140), Lowell (p.142), Maunder (p.143), Hall (p.144): Public domain / Sagan (p.145): NASA/JPL / Steltzner (p.145): NASA/Bill Ingalls / Zubrin (p.146): The Mars Society / Musk (p.147): Brian Solis / VR (pp149-150): NASA/James Blair / Earth (p.155): Shutterstock / Mars (p.155) Hellas Basin (p.96): MOLA Science Team